MATHEMATICS FOR COMMERCE

BASIC CONCEPTS, METHODS APPLICATION IN BUSINESS MATHEMATICS

I0465649

MATHEMATICS FOR COMMERCE

BASIC CONCEPTS, METHODS APPLICATION IN BUSINESS MATHEMATICS

K.SELVAKUMAR
M.Com,M.Phil,NET,Ph.D*

Notion Press

5 Muthu Kalathy Street, Triplicane,

Chennai - 600 005

First Published by Notion Press 2014

Copyright © K. Selvakumar 2014

All Rights Reserved.

ISBN: 978-93-84878-13-9

This book has been dedicated to students of
Andaman and Nicobar Islands

Preface

Mathematics is a subject enacted in the entire field like Science, Commerce and Arts. Knowledge of Mathematics also required our daily economical transaction like buying, selling, food preparing, tailoring, budgeting, etc. But, understanding of these concepts is not easier to everyone. It is one of the most important subjects which not only decides the careers of many young students but also enhances their ability of analytical and rational thinking. The knowledge about the methods, various formulae and regular practice only can make it easy and interesting. The students of Commerce and Economics in the Higher Education have been facing lot of difficulties to understand the concepts of Mathematics and its application. For long I have felt the need for a book incorporating the said features and containing a large number of well graded solved examples so that after studying it students can feel encouraged and develop confidence. Hence, an attempt has been made in this book to explain the basic concepts with easier way to understand every one. This book has been written as per the syllabus prescribed in U.G courses like B.Com (commerce), B.A (Economics), BBA under Pondicherry affiliated colleges for Business Mathematics. This book is the one of my first attempt. I develop this book on the base of the requirement of B.Com, B.A (Economics) Business Mathematics studying students under Pondicherry University. I hope this book will be very helpful to learn the basic Commercial and Econometrical Mathematics.

This book contains 8 units. The 1st Unit dealt about the **Ratio and proportion** used in business field and the 2nd Unit enumerate the **Profit and Loss** application in the Business. In continuation the 3rd Unit discussed about **Simple and Compound interest**. Shares, Debentures and Stocks are major part of the business transaction of joint stock companies. Unit 4 dealt about various aspects of **Shares, Debentures and Stock** like calculation of Interest, dividend, Market value etc. In many multinational corporation and Industries with large scale sectors are uses various management research technology. So, the basic concepts of **Functions and Relations**, **Matrices and determinants**, **Differentiation** and **Integration** has been explained in respective Units of 5th, 6th ,7th and 8th. Each unit took a special care about the basic concepts, formula and relevant illustrations arranged sequence of increasing the standard of questions. Each unit added special question for the students as problems for practice with answers. A list of formulae and University examination paper has been enclosed at the end. It is hoped that with all these unique features the book will pander to the needs of students.

K. SELVAKUMAR

Acknowledgement

I should great thankful to my parents who gave birth me and from the childhood they would me towards work culture, sincerity and honesty. Further, their love, blessing and encouragements may help to reach my destination.

I could not forget my beloved wife **Mrs. S. Vasanthi**, she helped lot physically and morally which would make very supportive to end this successfully. My special wishes to my son **M.S. Krishna Prabu** and my sweet daughter **S. Nandika Balashree.**

I should thankful to my **students, friends and professors of MGGC**, who has directly and indirectly help to publish this book successfully. Specially, I would like to thank **The Principal of Mahatma Gandhi Govt. College, Mayabunder** who has provided computer with internet facilities so that I could be able to publish the book timely.

I also very thankful to **Notion Press**, for their support to accept my first book for first edition which could motivate me for the future publications in this field of Business and management.

I shall be very much thankful to the **readers and users** for their support and suggestions for further improvement of my Book.

K. SELVAKUMAR

Contents

- ❖ Meaning of symmetric and skew-symmetric
- ❖ Determinants
- ❖ Properties of determinants
- ❖ Problems on determinants
- ❖ Ad joint of a matrix
- ❖ Inverse of a matrix
- ❖ Homogeneous system of linear equation

- ❖ Concept of limit and formula
- ❖ Differentiation meaning
- ❖ Rules of differentiation
- ❖ Addition and subtraction
- ❖ Multiple function and division function
- ❖ Exponential and logarithmic function
- ❖ Chain rule applicable in parametric function
- ❖ Trigonometric differentiation.

- ❖ Meaning
- ❖ Basic Rules of integration
- ❖ Some standard results
- ❖ Methods of integration
- ❖ Substitute method
- ❖ Partial fraction
- ❖ Integration by parts.

UNIT: I

RATIO AND PROPORTION

Meaning of ratio

Ratio is a numerical relation between two similar kinds.

A ratio is the relation which one quantity bear to another of the same kind, the comparison being made by containing what multiple parts of parts of II quantity is of the other. Commercial and Economically ratio and proportion making vital role in various calculation.

Uses of Ratios

Used in the analysis of financial statement by solepropriter, firm, company and finical institution to know their solvency and liquidity position.

It helps to reduce the larger quantities into very small so, that making analysis in all parameter is easier.

It is no units so, it reduce the confusion over the units and measurement of various commercial terms.

It is one of the easy techniques which were widely used to find various missing variables.

Types of Ratios

1. Duplicate ratio of a: b $= a^2 : b^2$

2. Sub duplicate ratio of a: b $= \sqrt{a} : \sqrt{b}$

3. Triplicate ratio of a: b $= a^3 : b^3$

4. Sub triplicate ratio of a:b $= \sqrt[3]{a} : \sqrt[3]{b}$

Q.1 converts the following into ratio:

a) 5 cm to 2m

Sol. $\dfrac{5cm}{2m}$

$= \dfrac{5cm}{2 \times 100cm}$

$= \dfrac{1}{40}$

$= 1:40$

b) 2m to 1.5 km

Sol. $\dfrac{2m}{1.5km}$

$= \dfrac{2m}{1500m}$

$= \dfrac{1}{750}$

$= 1:750$

Q.2 Given 2x+9y:3x+4y = 3:4, find the ratio of x to y

Sol. 2x+9y:3x+4y =3:4

$$\Rightarrow \frac{2x+9y}{3x+4y} = \frac{3}{4}$$

$$\Rightarrow 4(2x+9y) = 3(3x+4y)$$

$$\Rightarrow 8x+36y = 9x+12y$$

$$\Rightarrow 36y-12y = 9x-8x$$

$$\Rightarrow 24y = x$$

$$\Rightarrow \frac{24}{1} = \frac{x}{y}$$

$$\Rightarrow X: y = 24:1$$

Q.3 Given $3X+2y: 7x+y = 3:5$ find x: y

Sol. $3x+2y:7x+y=3:5$

$$\Rightarrow \frac{3x+2y}{7x+y} = \frac{3}{5}$$

$$\Rightarrow 5(3x+2y)=3(7x+y)$$

$$\Rightarrow 15x+10y=21x+3y$$

$$\Rightarrow 10y-3y=21x-15x$$

$$\Rightarrow 7y=6x$$

$$\Rightarrow \frac{x}{y} = \frac{7}{6}$$

$$\Rightarrow x: y=7:6$$

Q.4 Two numbers are in the ratio of 3:4. If 6 is added to the number the ratio of new number become 4:5 find the numbers.

Sol. Let, the number being x and y

X: y=3:4

$$\Rightarrow \frac{x}{y}=\frac{3}{4}$$

4x=3y,

$$y =\frac{4x}{3} \text{ .... (i)}$$

If 6 is added to each number then the new result

$$\Rightarrow \frac{x+6}{y+6} =\frac{4}{5}$$

$$\Rightarrow 5(x+y)=4(y+6)$$

$$\Rightarrow 5x+30=4y+24$$

$$\Rightarrow 5x+30=4(\frac{4x}{3})+24 \text{ (from (i)}$$

$$\Rightarrow 5x-\frac{16x}{3}=24-30$$

$$\Rightarrow \frac{-x}{3}=\frac{-6}{1}$$

$$\Rightarrow x=y$$

$$\Rightarrow y=\frac{4x}{3}$$

$$\Rightarrow y=\frac{4X18}{3}$$

$$\Rightarrow y=24$$

Q.5 Two observer find the ratio of some weight on different basis to be (2x-a): (3x-2a) and (2x+2a): (3x+2a), find 'x' and the value of ratio.

Sol. We know that ratio of same relation with different base are equal therefore

(2x-a): (3x-2a)=(2x+2a):(3x+2a)

$$\Rightarrow \frac{2x-a}{3x-2a}=\frac{2x+2a}{3x+2a}$$

$\Rightarrow$(2x-a) (3x+2a)=(2x+2a) (3x-2a)

$\Rightarrow$6x^2-4xa-3xa-2a^2 =6x^2-4ax+9ax-6a^2

$\Rightarrow$ax-2a^2=5ax-6a^2

$\Rightarrow$6a^2-2a^2=5ax-ax

$\Rightarrow$4a^2=4ax

Divide by 4a

$X = a$

Ratio of the number $= \dfrac{2x-a}{3x-2a}$

$$= \dfrac{2a-a}{3a-2a}$$

$$= \dfrac{a}{a}$$

$$= \dfrac{1}{1}$$

$$= 1{:}1$$

Q. 6 The monthly salary of two person in the ratio of 3:5, If each receive an increase in ₹ 20 in the monthly salary. The ratio is alter to 13:21 find their salaries.

Sol. Let the salary of the two person be ₹ 3x and ₹ 5x

By the condition we gat; a fraction form;

$$\Rightarrow \frac{3x+20}{5x+20} = \frac{13}{21}$$

$\Rightarrow 21(3x+20) = 13(5x+20)$

$\Rightarrow 63x+420=65x+260$

$\Rightarrow 420-260=65x-63x$

$\Rightarrow 160=2x$

$\Rightarrow x=\frac{160}{2}$

$\Rightarrow x=80$

Q.7 The ratio of price of two house 16:23, two years later the price of 1st one increase by 10% and the 2nd one increase by 477. The ratio of the prices becomes 11:20. Find the original price of the house.

Sol. Let, the original price of the house be = ₹ 16x and ₹ 23x

By the conditions we get the fraction;

$$\Rightarrow \frac{16x+(10\% \, of \, 16x)}{23x+477} = \frac{11}{20}$$

$$\Rightarrow \frac{16x+1.6x}{23x+477} = \frac{11}{20}$$

$$\Rightarrow \frac{17.6x}{23x+477} = \frac{11}{20}$$

$\Rightarrow 20 \times 17.6x = 11(23x+477)$

$\Rightarrow 352x = 253x+5247$

$\Rightarrow 352x-253x = 5247$

$\Rightarrow 99x = 5247$

$\Rightarrow x = \frac{5247}{99}$

$\Rightarrow x = 53$

Q. 8 Find the duplicate ratio of $2\sqrt{2} : 7\sqrt{3}$

Sol. We know that Duplicate ratio of $a: b = a^2:b^2$

Duplicate ratio of $2\sqrt{2} : 7\sqrt{3}$

$$= (2\sqrt{2})^2 : (7\sqrt{3})^2$$

$$= 8:174$$

Q. 9 Find the duplicate ratio of $5\sqrt{3} : 12\sqrt{2}$

Sol. We know that Duplicate ratio of $a:b = a^2:b^2$

$= (5\sqrt{3})^2 : (12\sqrt{2})^2$

$= 75:288$

Q. 10 find the sub. Duplicate ratio of the following

a) 48:124

Sol. We know that sub-duplicate ratio of $a:b = \sqrt{a} : \sqrt{b}$

48:124

$= \sqrt{48} : \sqrt{124}$

$= 4\sqrt{3} : 2\sqrt{31}$

$= 2\sqrt{3} : \sqrt{31}$

b) 128 : 256

Sol. We know that sub-duplicate ratio of a:b $= \sqrt{a} : \sqrt{b}$

So, sub-duplicate ratio of 128 : 256

$= \sqrt{128} : \sqrt{256}$

$= 8\sqrt{2} : 16$

TRIPLICATE RATIO

Q. 11 find the triplicate ratio of the following:

a) $2\sqrt{3} : 3\sqrt{2}$

Sol. The triplication of ratio is a:b $= (a)^3 : (b)^3$

So, Triplicate ratio of $2\sqrt{3} : 3\sqrt{2}$

$= (2\sqrt{3})^3 : (3\sqrt{2})^3$

$= 8(\sqrt{3})^3 : 27(\sqrt{2})^3$

$= 8 \times 3\sqrt{3} : 27 \times 2\sqrt{2}$

$= 24\sqrt{3} : 54\sqrt{2}$

$= 4\sqrt{3} : 9\sqrt{2}$

b) $7\sqrt{5} : 3\sqrt{7}$

Sol. The triplicate ratio of a:b $= (a)^3 : (b)^3$

$\therefore$ Triplicate ratio of7 $\sqrt{5}$:3 $\sqrt{7}$

$=(7\sqrt{5})^3 : (3\sqrt{7})^3$

$=343X\sqrt{5} : 27X\sqrt{7}$

$=49X\sqrt{5} : 27\sqrt{7}$

$=245\sqrt{5} : 27\sqrt{7}$

SUB. TRIPLICATE RATIO

Q. Find the sub-triplicate ratio of the following.

a) 125:345

Sol. The sub- triplicate ratio are a:b $= \sqrt[3]{a} : \sqrt[3]{b}$

$= (a)^{1/3} : (b)^{1/3} = (125)^{1/3} : (343)^{1/3}$

$= (5^3)^{1/3} : (7^3)^{1/3}$

$= (5)^{3\times 1/3} : (7)^{3\times 1/3}$

$= 5:7$

b) 729:512

Sol. The sub- triplicate ratio are a:b $= \sqrt[3]{a} : \sqrt[3]{b}$

$=(a)^{1/3} : (b)^{1/3}$

$= (729)^{1/3} : (512)^{1/3}$

$$=(9^3)^{x1/3} : (8^3)^{1/3}$$

$$=(9)^{3x1/3} : (8)^{3x1/3}$$

$$=9:8$$

PROBLEMS FOR PRACTICE

1. Convert into ratio in smallest terms

1) ₹ 5 to 200 Paisa

2) 5m to 2.5m

3) 200g to 2kg

4) 50ml to 1l

5) 3months to 2 years

6) 50seconds to 2 minutes

7) 15minutes to 1 hours

8) Two numbers are in the ratio of 1:3. If 15 is added into the both number the ratio become 4:9.Find the original numbers.

9) The Price of two commodities is the ratio of 16:25. If the price is increased by ₹20000 each the new price becomes 17:26.Find the cost of each commodity.

10) Find x:y ,If (10x+3y): (5x+2y)= 9:5.

11) The monthly income of two persons is in the ratio of 6:7 and their monthly expenditure is the ratio of 11:13, if each one saves 50. Find their monthly income and expenditure.

12) The number is in the ratio of 3:4.If 6 is added to the number the ratio of new number become 4:5. Find the numbers.

13) Find the duplicate Ratio of the following a) 7:9 b) $11 \sqrt{2} : 13 \sqrt{3}$

14) Find sub-duplicate ratio of a) 50:128 b) 27:125

15) Find Triplicate Ration of a) $5 \sqrt{2} : 3 \sqrt{5}$ b) ($\sqrt{2}$ +1): ($\sqrt{2}$ -1)

16) Find sub-Triplicate ratio of a) 1000:729 b) 128:676

Answers: 1)5:2 2) 2:1 3)1:10 4)1:20 5)1:8 6)5:12 7)1:4 8)25 and75 9) ₹ 320000 and ₹500000 10) x: y= 5:3 11) Income ₹600 and ₹ 700, Expenditure ₹550 and ₹650 12) 18 and 24 13) a. 49:81 b.242:507 14) a) $5 \sqrt{2}$:8 $\sqrt{2}$ b) $3 \sqrt{3}$:5 $\sqrt{5}$ 15) a) 50 $\sqrt{2}$: 27 $\sqrt{5}$ b) 1:1 16) a) 10:9 b) $4\sqrt[3]{2} : \sqrt[3]{676}$

PROPORATION

When two ratio a: b and c: d is equal then they are said to be in proportion. So we can define the proportion as, 'Proportion expresses equality between two ratios'.

If $\frac{a}{b} = \frac{c}{d}$ (on a: b = c: d then a&d are extremes and b&c are called means)

Many mathematical relations can be expressed in proportion and it is widely used to find many business variables. For estimating, budgeting by using the related variables can be easily assessable under this proportion method.

PROPERTIES OF PROPORTION

Fourth proportion to a, b and c is i.e. x then a: b = c: x

Third proportion to 'a' and 'b' is x then a: b = b: x

Mean proportion to 'a' and 'b' is x then a: x = x: b

If a, b, c and d are continued proportion then $\frac{a}{b} = \frac{b}{c} = \frac{c}{d}$

If $\frac{a}{b} = \frac{c}{d}$ by the alternando $\Rightarrow \frac{a}{c} = \frac{b}{d}$

If $\frac{a}{b} = \frac{c}{d}$ by then invertendo $\Rightarrow \frac{b}{a} = \frac{d}{c}$

If $\frac{a}{b} = \frac{c}{d}$, by the componendo $\Rightarrow \frac{a+b}{b} = \frac{c+d}{d}$

If $\frac{a}{b} = \frac{c}{d}$, by Divindendo $\Rightarrow \frac{a-b}{b} = \frac{c-d}{d}$

If $\frac{a}{b} = \frac{c}{d}$, by compenendo & Dividendo $\Rightarrow \frac{a+b}{a-b} = \frac{c+d}{c-d}$

Q. Find the fourth proportion to 10, 13, 25

Sol. Let the following properties be 'x'

Therefore $10:13 = 25: x$

$\Rightarrow \frac{10}{13} = \frac{25}{x}$

$\Rightarrow 10x = 25 \times 13$

$\Rightarrow x = \frac{325}{10}$

$\Rightarrow x = 32.5$

b) 12, 96 and 112

Sol. Let the fourth properties be 'x'

$\Rightarrow \frac{12}{96} = \frac{112}{x}$

$\Rightarrow 12x = 112 \times 96$

$\Rightarrow 12x = 10,752$

$\Rightarrow x = \frac{10752}{12}$

$\Rightarrow x = 896$

Q.2 Find the third properties to the numbers 8 and 12.

a) Let the third properties be 'x'

$\Rightarrow \frac{8}{12} = \frac{12}{x}$

$\Rightarrow 8x = 12 \times 12$

$\Rightarrow 8x = 144$

$\Rightarrow x = \frac{144}{8}$

$\Rightarrow x = 18$

b) 15 and 115

Sol. Let the third proportion be 'x'

$15 : 115 = 115 : x$

$15/115 = 115/x$

$\Rightarrow 15x = 115 \times 115$

$\Rightarrow 15X = 13225$

$\Rightarrow x = 881.6$

Q.3 Find the mean proportion to the following:

a) 45 and 15

Sol. Let the mean proportion be x

$\Rightarrow \frac{45}{x} = \frac{x}{15}$

$\Rightarrow x^2 = 45 \times 15$

$\Rightarrow x^2 = 675$

$\Rightarrow x = \sqrt{675}$

$\Rightarrow x = 15\ \sqrt{3}$

b) 16 and 128

Sol. $\frac{16}{x} = \frac{x}{128}$

$\Rightarrow x^2 = 128 \times 16$

$\Rightarrow x^2 = 2048$

$\Rightarrow x = \sqrt{2048}$

$\Rightarrow x = 32\sqrt{2}$

Q.4 If a: b = c:d i.e. is equal 2.5:1.5,then find (i) ad:bc (ii) (a+c):(b+d)

$\Rightarrow \frac{a}{b} = \frac{c}{d} = \frac{2.5}{1.5} = \frac{25}{15} = \frac{5}{3}$

i) $\frac{a}{b} = \frac{c}{d}$ (Given)

$\Rightarrow ad = bc$

$\Rightarrow \frac{ad}{bc} = \frac{1}{1}$

$\Rightarrow$ ad: bc = 1:1

ii) $\frac{a}{b} = \frac{c}{d}$ (Given)

$\Rightarrow \frac{a+c}{b+d} = \frac{a}{b} = \frac{c}{d} = \frac{5}{3}$

$\Rightarrow$ a+b:b+d = 5:3

Q.5 If $\frac{a}{3} = \frac{b}{4} = \frac{c}{7}$ show that $\frac{a+b+c}{c} = 2$

Sol. We know that

$$\Rightarrow \frac{a}{b} = \frac{c}{d} = \frac{e}{f} = \frac{a+b+e}{b+d+f}$$

$$\Rightarrow \frac{a}{3} + \frac{b}{4} + \frac{c}{7} = \frac{a+b+c}{3+4+7} = \frac{a+b+c}{14}$$

$$\Rightarrow \frac{a+b+c}{14} = \frac{c}{7}$$

By the alternendo

$$\Rightarrow \frac{a+b+c}{c} = \frac{14}{7}$$

$$\Rightarrow \frac{a+b+c}{c} = 2$$

Q.6 If a+b: $\sqrt{ab}$ = 4:1 , then find a:b

Sol. a+b: $\sqrt{ab}$ = 4:1(given)

$$\Rightarrow \frac{a+b}{\sqrt{ab}} = \frac{4}{1}$$

$$\Rightarrow \frac{a}{\sqrt{ab}} + \frac{b}{\sqrt{ab}} = \frac{4}{1}$$

$$\Rightarrow \frac{\sqrt{a}}{\sqrt{b}} + \frac{\sqrt{b}}{\sqrt{a}} = 4$$

Again $\frac{a+b}{\sqrt{ab}} = \frac{4}{1}$

Multiply by $\frac{1}{2}$ for both sides

$$\Rightarrow \frac{a+b}{2\sqrt{ab}} = \frac{4}{2}$$

Applying 'c' & 'D'

$$\Rightarrow \frac{a+b+2\sqrt{ab}}{a+b-2\sqrt{ab}} = \frac{4+2}{4-2}$$

$$\Rightarrow \frac{(\sqrt{a}+\sqrt{b})^2}{(\sqrt{a}-\sqrt{b})^2} = \frac{6}{2}$$

$$\Rightarrow \frac{(\sqrt{a}+\sqrt{b})^2}{(\sqrt{a}-\sqrt{b})^2} = \frac{3}{1}$$

$$\Rightarrow \frac{\sqrt{a}+\sqrt{b}}{\sqrt{a}-\sqrt{b}} = \frac{\sqrt{3}}{1}$$

Applying C and D

$$\Rightarrow \frac{\sqrt{a}+\sqrt{b}+\sqrt{a}-\sqrt{b}}{\sqrt{a}+\sqrt{b}+\sqrt{a}+\sqrt{b}} = \frac{\sqrt{3}+1}{\sqrt{3}-1}$$

$$\Rightarrow \left(\frac{\sqrt{a}}{\sqrt{b}}\right)^2 = \left(\frac{\sqrt{3}+1}{\sqrt{3}-1}\right)^2$$

$$\Rightarrow \frac{a}{b} = \frac{3+2\sqrt{3}+1}{3-2\sqrt{3}+1}$$

$$\Rightarrow \frac{a}{b} = \frac{2+\sqrt{3}}{2-\sqrt{3}}$$

Q. 7 If $\frac{x}{5} = \frac{y}{6} = \frac{z}{7}$, find the value of $\frac{3x-2y+6z}{8x-5y-3z}$

Sol. $\frac{2}{5} = \frac{y}{6} = \frac{z}{7}$ (Given)

X=5k; y=6k; z=7k

$$\Rightarrow \frac{3x-2y+6z}{8x-5y-3z} = \frac{3(5k)-2(6k)+6(7k)}{8(5k)-5(6k)-3(7k)}$$

$$= \frac{15k-12k+42k}{40k+30k-21k}$$

$$= \frac{45k}{49k}$$

$$= \frac{45}{49}$$

Q. 8 If $\dfrac{x}{y+z} = \dfrac{y}{z+x} = \dfrac{z}{x+y}$ then prove that, if x+y+z=0, then each ratio $= \dfrac{1}{2}$

Sol. $\dfrac{x}{y+z} = \dfrac{y}{z+x} = \dfrac{z}{x+y}$

$\Rightarrow \dfrac{x+y+z}{y+z+z+x+x+y}$

$\Rightarrow \dfrac{x+y+z}{2x+2y+2z}$

$\Rightarrow \dfrac{(x+y+z)}{2(x+y+z)} \Rightarrow \dfrac{1}{2}$

Q. 9 If $\dfrac{a}{b} = \dfrac{c}{d}$, show that $\dfrac{4a+9b}{4a-9b} = \dfrac{4c+9d}{4c+9d}$

Sol. $\dfrac{a}{b} = \dfrac{c}{d}$

Multiply by $\dfrac{4}{9}$ for both sides

$\Rightarrow \dfrac{4a}{9b} = \dfrac{4c}{9d}$

$\Rightarrow \dfrac{4a+9b}{4a-9b} = \dfrac{4c+9d}{4c-9d}$

Q. 10 If $\dfrac{9x+7y}{9x-7y} = \dfrac{9p+7q}{9p-7q}$ show that x: y = p: q

Sol. Given $\dfrac{9x+7y}{9x-7y} = \dfrac{9p+7q}{9p-7q}$

Applying 'c' &'d'

$\Rightarrow \dfrac{9x+7y+9x-7y}{9x+7y-9x+7y} = \dfrac{9p+7q+9p-7y}{9p+7q-9p+7q}$

$\Rightarrow \dfrac{18x}{14y} = \dfrac{18p}{14q}$

Divide by $\frac{18}{14}$ for the both side

$\Rightarrow \frac{x}{y} = \frac{p}{q}$

$\Rightarrow$x: y = p: q

Q.11 Using properties of proportion solve for x, $\frac{\sqrt{x+2}}{\sqrt{x+2}} + \frac{\sqrt{x-3}}{\sqrt{x-3}} = 5$

Sol. Given $\frac{\sqrt{x+2}}{\sqrt{x+3}} + \frac{\sqrt{x-3}}{\sqrt{x-3}} = 5$

Applying 'c' &'d'

$\frac{\sqrt{x+2}+\sqrt{x-3}+\sqrt{x+2}-\sqrt{x-3}}{\sqrt{x+2}+\sqrt{x-3}-\sqrt{x+2}+\sqrt{x-3}} = \frac{5+1}{5-1}$

$\frac{2\sqrt{x+2}}{2\sqrt{x-3}} = \frac{6}{4}$

Squaring both side

$\frac{x+2}{x-3} = \frac{9}{4}$

9(x-3) =4(x+2)

9x-27 =4x+8

5x=35

X=5

Q. 12 $\dfrac{\sqrt{x+4}+\sqrt{x-10}}{\sqrt{x+4}-\sqrt{x-10}} = \dfrac{5}{2}$, find x.

Sol. Given $\dfrac{\sqrt{x+4}+\sqrt{x-10}}{\sqrt{x+4}-\sqrt{x-10}} = \dfrac{5}{3}$

Apply 'c' &'d'

$$\Rightarrow \frac{\sqrt{x+4}+\sqrt{x-10}+\sqrt{x+4}+\sqrt{x-10}}{\sqrt{x+4}-\sqrt{x-10}-\sqrt{x+4}-\sqrt{x-10}} = \frac{5+2}{5-2}$$

$$\Rightarrow \frac{\sqrt{x+4}+\sqrt{x+4}}{\sqrt{x-10}+\sqrt{x-10}} = \frac{7}{3}$$

$$\Rightarrow \frac{2\sqrt{x+4}}{2\sqrt{x-10}} = \frac{7}{3}$$

Squaring both side

$$\Rightarrow \left(\frac{\sqrt{x+4}}{\sqrt{x-10}}\right)^2 = \left(\frac{7}{3}\right)^2$$

$$\Rightarrow \frac{x+4}{x-10} = \frac{49}{9}$$

$$\Rightarrow 49(x-10) = 9(x+4)$$

$$\Rightarrow 49x-490 = 9x+36$$

$$\Rightarrow 49x-4x = 36+490$$

$$\Rightarrow 40x = 526$$

$$\Rightarrow x = \frac{526}{40}$$

$$\Rightarrow x = 13.15$$

Q. 13 A trader mixes two kinds of tea leaves which cost him ₹ 24 & ₹ 18 per kg 98p. A mixing tea 1kg in every 100kg is wasted. In what proportion he must to mix them, the cost per kg become ₹ 20.

Sol. Let the mixtures of tea be 'x' kg from 1st item and 'y' kg from 2nd item

$$= x \times 24 + yX18 = (x+y) \ x20x\frac{99}{100}$$

$$\Rightarrow 5(24x+18y) = 99x+99y$$

$$\Rightarrow 120x+90y = 99x+99y$$

$$\Rightarrow 120x-99x = 99y-90y$$

$$\Rightarrow 21x = 9y$$

$$\Rightarrow \frac{x}{y} = \frac{9}{21}$$

$$\Rightarrow \frac{x}{y} = \frac{3}{7} \Rightarrow x: y = 3:7$$

Q.14 A trader mixed two kinds of tea in the ratio of 2:1 and a profit of 25% by selling the mixtures @ ₹ 70 per the makes profit of the same ratio on mixing them in the ratio 3:2 and selling the mixtures @ ₹ 72 per kg. Find the profit which he bought them.

Sol. Let, the price of 1st item of tea per kg = ₹ X

The price of 2nd item of tea per kg = ₹ Y

According to the condition

$$(2x + 1y) \ x\frac{125}{100} = 70x3$$

$$\Rightarrow \frac{5}{4}(2x+y)= 210$$

$$\Rightarrow 2x+y = \frac{210}{5}x4$$

$$\Rightarrow 2x+y= 168 \dots\dots\dots (i)$$

Again $(3x+2y) \, X\frac{125}{100} = 72X5$

$$\Rightarrow 3x+2y = 288 \dots\dots\dots (ii)$$

Solving equation (i) & (ii)

$$\Rightarrow 2x+y=168$$

$$3x+ 2y=288$$

Multiply equation (i) by 2

$$4x + 2y = 336$$
$$3x + 2y = 288$$

$$X = 48$$

When, x = 48, y =?

$$2(48) +y = 168$$

$$Y=168-96$$

$$Y=72$$

Q. 15 If $\frac{a}{b} = \frac{c}{d} = \frac{e}{f}$ show that; $\sqrt{ab} + \sqrt{cd} + \sqrt{ef} = \sqrt{(a + e + e)(b + d + f)}$

Sol. $\frac{a}{b} = \frac{c}{d} = \frac{e}{f} = k$

$\Rightarrow$a=bk, c=dk, e=fk

L.H.S=$\sqrt{ab} + \sqrt{cd} + \sqrt{ef}$

$=\sqrt{(bk)(b)} + \sqrt{(dk)d} + \sqrt{(fk)f}$

$= \sqrt{b^2 k} + \sqrt{d^2 k} + \sqrt{f^2 k}$

$= b\sqrt{k} + d\sqrt{k} + f\sqrt{k}$

$= (b + d + f) \sqrt{k}$

R.H.S $= \sqrt{(a + c + e)(b + d + f)}$

$= \sqrt{(bk + dk + fk)(b + d + f)}$

$= \sqrt{k(b + d + f)(b + d + f)}$

$= \sqrt{k(b + d + f)^2}$

$= (b + d + f) \sqrt{k}$

L.H.S = R.H.S

PROBLEMS FOR PRACTICE

1) Find fourth proportion to a) 112, 200 and 512 b) 78,128 and 576 (**Ans: a) 914.3 b) 945.2**

2) Find third proportion to a) 6 and 7 b) 12 and 38 (**Ans: a) 8.2 b) 120.3**)

3) Find mean proportion to a) 20 and 80 b) 12 and 48 (Ans: **a) 40 b)24**)

4) If $\frac{a}{4} = \frac{b}{5} = \frac{c}{6}$. Find the value of $\frac{a+b+c}{c}$ (Ans: **2:5**)

5) Two numbers are in the ratio of 1:5.If 5 is added to the number the new number are become 3:11. Find the numbers. (**Ans: 10**)

6) What must be added to 10, 18, 22 and 38 so that they become in proportion (**Ans: 2**)

7) If $\frac{a}{b} = \frac{c}{d}$, show that 6a+ 7b:6c+7d = 6a-7b: 6c-7d.

8) Solve of x. $\frac{\sqrt{x+7} + \sqrt{x-1}}{\sqrt{x+7} + \sqrt{x-1}} = \frac{2}{1}$ (**Ans:2**)

9) If $\frac{a}{b} = \frac{c}{d}$, prove that $\frac{a^2+b^2}{ab+cd} = \frac{a^2+b^2}{b^2+d^2}$,

10) If $\frac{a}{b} = \frac{c}{d} = \frac{e}{f}$. Prove that $(ab+cd+ef)^2 = (a^2+c^2+e^2)(b^2+d^2+f^2)$

CONTINUED PROPORTION

If a, b, c and d are in continued proportion then we have $\frac{a}{b} = \frac{b}{c} = \frac{c}{d}$ .We can use K method to solve the question which based on continued proportion.

Q.1 If a b and c are continued proportion a: c = (a² + b²) :(b²+c²)

Sol. If a b and c are in continued proportion then we have $\frac{a}{b} = \frac{b}{c}$ =k

a=bk, b=ck

a=ck², b=ck

L.H.S = a:c

$$=\frac{a}{c}=\frac{ck^2}{c}$$

$$=k^2$$

R.H.S $= (a^2+b^2):(b^2+c^2)$

$$=\frac{a^2+b^2}{b^2+c^2}$$

$$=\frac{c^2k^4+c^2k^2}{c^2k^2+c^2}=k^2$$

Q.2 If a, b, c, d are continued, prove that $\dfrac{a}{b}=\dfrac{a^3+b^3+c^3}{b^3+c^3+d^3}$

Sol. Given a, b, c, and d are continued proportion

$$=\frac{a}{b}=\frac{b}{c}=\frac{c}{d}=\frac{k}{1}$$

a=bk, b=ck, c=dk

a=dk³, b=dk², c=dk

L.H.S $=\dfrac{a}{b}$

$$=\frac{dk^3}{d}$$

$$=k^3$$

R.H.S $=\dfrac{a^3+b^3+c^3}{b^3+c^3+d^3}$

$$=\frac{(dk^3)^3+(dk^2)^3+(dk)^3}{(dk^2)^3+(dk)^3+d^3}$$

$$=\frac{d^3k^3\left(k+k^3+1\right)}{d^3(k+k^3+1)}$$

$$=k^3$$

L.H.S = R.H.S

PROBLEMS FOR PRACTICE

1. If $\frac{a}{b} = \frac{c}{d}$ .Prove that $a+b : c+d = \sqrt{a^2 + b^2} : \sqrt{c^2 + d^2}$.

2. If a, b, and c are in continued proportion .Prove that $(a+b+c)(a-b+c) = a^2+b^2+c^2$

3. If a, b,c and d are in continued proportion .prove that

    i) $\frac{2a+b}{2a+c} = \frac{2b+c}{2c+d}$ ii) $\frac{a^2+b^2+c^2}{ab+bc+cd} = \frac{ab+bc+cd}{b^2+c^2+d^2}$

DIRECT PROPORTION (VARIATION)

If two quantities are said to varied directly (direct proportion) if they increase or decrease in one of the quantity causes the increase or decrease in other quantity.

For examples. The cost of articles and number of articles.

The distance over by moving object where is directly as its speed

The work done varies directly as the number of man at work and the work done varies directly as the working time.

Q. 1 if ₹ 166.50 is the cost of 9kg sugar. How much sugar can be purchase for ₹ 259?

Sol. Here, the quantity of sugar and cost of sugar are direct variation let x kg of sugar can be brought from ₹259.

Quantity of sugar (kg)	9	X
Cost of sugar	166.50	259

$$\Rightarrow \frac{9}{166.50} = \frac{x}{259}$$

$$\Rightarrow x \times 166.50 = 9 \times 259$$

$$\Rightarrow x = \frac{2331}{166.50}$$

$$\Rightarrow x = 14$$

Q.2 If one score orange cost ₹45. How many oranges can be bought for ₹72?

Sol. Let number of orange denote x and cost per oranges ₹ Y. according to the questions

X	20	X
Y	45	72

X and y varied directly

$$\Rightarrow \frac{20}{45} = \frac{x}{72}$$

$$\Rightarrow 45 \times x = 72 \times 20$$

$$\Rightarrow x = \frac{72 \times 20}{45}$$

$$\Rightarrow x = \frac{1440}{45}$$

$\Rightarrow x=32$

Q.3 If a car cover 82.5 kilometer in 5.5 litters petrol. How much distance will it cover in 13.2 litres of petrol?

Quantity of petrol	5.5	13.2
Distance (km)	82.5	Y

There, using of petrol and distance travelled are direct variation

$$\Rightarrow \frac{5.5}{82.5} = \frac{13.2}{y}$$

$\Rightarrow 5.5xy=132X82.5$

$$\Rightarrow y=\frac{13.2X82.5}{5.5}$$

$$\Rightarrow y=\frac{1089}{5.5}$$

$\Rightarrow y=198$

Q.4 If 5 men or 7 women can earn ₹875 per day. How much income 10 men and 5 women earn per day?

Sol. Let change the relation to woman.

5 men $\quad$ = 7 women

1 man $\quad$ = $\frac{7}{5}$ women

10 man $\quad$ = $\frac{7}{5}$x100 women

$\qquad\qquad$ =14 women

10 men and 5 women = 14+5=19 women

No. of women	7	19
Earning (7)	875	y

$$\Rightarrow \frac{7}{875} = \frac{19}{y}$$

$$\Rightarrow 7xy = 19 \times 875$$

$$\Rightarrow y = \frac{19 \times 875}{7}$$

$$\Rightarrow y = 2375$$

INVERSE VARIATION

Two quantities are such to vary inversely (inverse variation) if the increase or decrease in one quantity cause the decrease or increase in the other quantity for example:

i. Time taken to finished a piece of work varies inversely as the number of workers at work (more man less time, less man more time)

ii. Speed varies inversely as a time taken to cover the distance (more speed less time, less speed more time)

iii. Demand of a product in respect of price is also inverse relation.

Q. 1 35 men can reap a field in 8 days, how many days can 20 men reap the same field.

Sol.

No. of man	35	20
No. of days	8	Y

Here the no. of man and the time taken are inversely varied

$\Rightarrow 35 \times 8 = 20 \times y$

$\Rightarrow y = \dfrac{30 \times 8}{20}$

Therefore 20 men can finish it in 14 days.

Q.2 A food head a provision for 300 men for 90 days after 20 days 50 men left from the part, how long would the food last at the same rate.

Sol. Remaining number of man = 300-50 = 250

Remaining days left = 90-20=70

Here the food provision and number of person varied inversely i.e., less man more days.

No. of man	300	250
No. of days	70	Y

$300 \times 70 = 250 \times y$

$Y = \dfrac{300 \times 70}{250} = 84$

Q.3 6 oxen or 8 cows can graze a field in 28 days. How long would 9 0xen and two cows take to graze to the same field.

Sol. We know,

6 oxen = 8 cows

1 ox = $\frac{8}{6}$ cows

9 oxen and 2 cows = 9 X $\left(\frac{8}{6}\right)$ cows + 2 cows

= 12+2 = 14 cows

No. of cows grazed	8	14
No. of day last	28	y

$\Rightarrow$ 8X28=14Xy

$\Rightarrow y=\frac{8X28}{14}$

$\Rightarrow y=\frac{224}{14}$

$\Rightarrow y=16$

Q.4 12 men and 15 women can finished a piece of work in 66 days. How long 24 men and 3 women take to finish the work?

Sol. We know,

12 men = 15 women

1man = $\frac{15}{12}$ women

24 men and 3 woman = 24 X $\frac{15}{12}$ men + 3 women

= 30+3

= 33 women

No. of woman worked	15	33
No. of days	66	y

$\Rightarrow 15 \times 66 = 33 \times y$

$\Rightarrow y = \frac{15 \times 66}{33}$

$\Rightarrow y = 30$

Q.5. If 600 men can dig a trench 5.5m broad, 4 m and 405m long if half an hour. What length in kms of a trench 10m broad and 8m deep can 25000 men in 6 hrs.

Sol. The problem is based on compound proportion so we can rearrange in the following form

Men: 600:25000

Broad 10: 5.5

Deep: 8: 4 :: 405: x

Hours: $\frac{1}{2}$: 6

i.e., $600 \times 10 \times 8 \times \frac{1}{2} \times x = 25000 \times 5.5 \times 4 \times 6 \times 405$

$x = \frac{25000 \times 5.5 \times 4 \times 6 \times 405 \times 2}{600 \times 10 \times 8} = 55.68$

So, they can dig 55.68 Km (approx).

PROBLEMS FOR PRACTICE

1. 18dolls cost ₹630.How much dolls can be brought ₹ 455. **(Ans: ₹ 45)**

2. If 5 men or 7 women can earn ₹525, how much 7 men and 13 women earn for day. **(Ans:₹1725)**

3. 12 men can dig a pond in 8 days. How many men can dig it in 6 days.**(Ans-16 men)**

4. A fort had enough food for 80 soldiers for 60 days. How long would the food last if 20 more soldier joint after 15 days.**(Ans:36days)**

5. If the wages of 12 workers for 5 days are ₹ 7500.Find the wages of 17 workers for 6 days.**(Ans: ₹12750)**

PERCENTAGE

The term percentage means out of hundred. If any numerical terms are expressed in terms of hundred is known as percentage and it is denoted (%). It is used to reduced the very complication figures within hundred and we predict many commercial term in percentage for easier understand. It is very useful to find % of profit of businessmen, firm and other commercial sectors.

In financial institution, commercial banks the use of % is very essentials to calculate interest to the investors and borrowers. Statistical data can be summarized by using the techniques of percentage. For the financial statement analysis the technique of % are useful to find various profitability ratios.

Q.1 Convert the following into (%)

a) $\frac{5}{8} \Rightarrow \frac{5}{8} \times 100$

$\Rightarrow 62.5\%$

b) $\frac{25}{125} = \frac{25}{125} \times 100$

$= 20\%$

C) $0.25 = 0.25 \times 100$

$= 25\%$

Q.2 Find a) $2\frac{1}{2}\%$ of 250

Sol. $2\frac{1}{2}\%$ of 250

$= \frac{5}{200} \times 250 = 6.25$

b) Find 'x' if 7.5% of x is ₹ 5000

Sol. 7.5% of X = 5000

$\Rightarrow x = \frac{5000 \times 100}{7.5}$

$\Rightarrow x = 66,666$

Q.3 A water tank contain tank 10,000 l water if 500 l of water get leakage in two days, find the percentage of water get leakage and % of water left on tank.

Sol. Capacity of water in the tank = 10000 l

Capacity of water leakage = 500 l

Percentage of water leakage in the tank $= \frac{500}{10000}$ X 100

=5%

Percentage of water left in bank = 100-5 = 95%

Q. 4 A person get a monthly salary of ₹ 25000. If his salary is increase at 5% what it his new monthly salary?

Sol. Total salary of a person = ₹ 25000

Percentage of salary increase = 5%

Percentage of new monthly salary is increase = ₹ 25000 X 5%

$= 1,250$

New monthly salary = 25000+1250 =₹ 26250

Q. 5 Mr. Ramesh score marks out of 100 in sixth subjects are 67, 72, 93, 48, 55 & 87. Find his percentage of marks score.

Sol. Total marks =600

Obtained marks =67+72+93+48+55+87=422

% of marks score $=\frac{422}{600}$ X 100

$= 70\%$

Q. 6 Find the value of x. if 55% of 80+30% of (90+x) =198

Sol. 55% of 80 + 30% of (90+X) =198

$$\Rightarrow \frac{55}{100} X80 + \frac{30}{100} X 90 + x = 198$$

$$\Rightarrow 44+27+x=198$$

$$\Rightarrow x=198-71$$

$$\Rightarrow x=127$$

Q.6 Gullson monthly income is ₹ 1800. If he spends 15% of food 15% on cloth 10% on house rent and 50% on domestic use. What is his saving?

Sol. Percentage of saving = 100-(15+15+10+50)

=100-90

= 10%

Amount of saving = 10% of 1800

$$= \frac{10}{100} X 1800 = 180$$

Q.7 The population of a city was 50000. It is increase to 52000. What is the percentage of increase in population?

Sol. Initial population = 50000

Present population = 52000

Increase in population = 52000-50000

=2000

Percentage of increase in population $= \frac{2000}{50000} X 100$

=4%

PROBLEMS FOR PRACTICE

1. Convert the following into % a) 8/12 b) 0.128 c) $5\frac{1}{4}$

2. Find the value of x: if $5\frac{1}{8}$ % of x is 28.

3. Which is greater 2% of 28 or 15% of 12?

4. If Mr. Ravi received ₹550 wages offer an increment of11% from his initial wages. What is his initial wages?

5. Narendra deposits 15% of his salary in the fixed deposit account and after spending 30% of the remainder is left with ₹2380. What will be his salary?

6. In an examination 45% students failed in history and 30% students failed in English. If 15% students failed in both subjects then what is the pass percentage of the total students, passed in both subjects.

7. A house is injured for $\frac{4}{5}$ of its original value. If the amount premium at the rate of 1.3% is ₹910, then what is the original value of that house?

8. If A's income be 10% more than B's income, how much % is the income less than A?

Ans 1. A) 66.6% b) 0.128% c) 525 % 2)546 3) 15% of 12 4) ₹ 500 5) ₹4000 6)45% 7)87500 8)$9\frac{1}{11}$ %

UNIT:II

PROFIT AND LOSS

FORMULA

Profit of Gain = sp – cp

Loss = cp-sp

Sp (on gain) = cp $(\frac{100+p\%}{100})$

Sp (on loss) = cp $(\frac{100-l\%}{100})$

P% = $\frac{p}{cp}$ X 100

L% = $\frac{L}{CP}$ X100

Discount is allowed on basic price (or) market price (or) list price.

Sp (on discount) = mp $(\frac{100-d\%}{100})$

D% = $\frac{Discount}{mp}$ X100

Q.1 Mohit bought a CD of ₹ 750 and sold it in ₹ 875. Find gain or loss percent.

Sol.   Sp of C.D = 875

Cp of C.D = 750

Profit = Sp – Cp

$= 875 - 750$

$= ₹\ 125$

% of profit $= \frac{p}{cp}$ x 100

Q.2 Mohan Lal purchase on old scooter for ₹ 12000 and spend ₹ 2850 on his overhauling then he sold it to his friend for ₹ 13860. How much % did him gain or loss.

Sol. List price of scooter = purchase price + overhauling expenses

$$= ₹\ 12000 + ₹\ 2850$$

$$= ₹\ 14850$$

Sp of scooter = Cp – sp

$$=₹\ 14850 - 13860$$

$$=990\%$$

% of loss $= \frac{l}{cp}$ X 100

$$=\frac{990}{14850}\ X\ 100$$

$$=6.67\%$$

Q. 3 A vendor bought orange at 20 for ₹ 56 and sold them ₹ 35 per dozen. Find profit or loss %.

Sol. Cp of 20 oranges = ₹ 56

Sp of 12 oranges = ₹ 35

$\therefore$ 1 orange $= \dfrac{35}{12}$

$\therefore$for 20 oranges $= \dfrac{35}{12}$ x20

$\qquad\qquad = ₹58.33$

Profit $=$ Sp-Cp

$\qquad = 58.33\text{-}56.00$

$\qquad = ₹2.33$

Profit % $= \dfrac{P}{cp}$ x100

$\qquad\quad = \dfrac{2.33}{56}$ x100

$\qquad\quad = 4.6\%$

Q.4 if the cost price of 10 greeting cards is equal to selling price of 8 greeting cards. Find the gain or loss %.

Sol. Let sp of 1 greeting card $= ₹$ x

Cp of 10 greeting cards $= ₹$ 8x

Sp of 10 greeting cards $= 10$x

Profit $=$ Sp- Cp

$\qquad\quad = 10\text{x - }8\text{x}$

$\qquad\quad = 2\text{x}$

Profit % $= \dfrac{P}{cp}$ x100

$$= \frac{2x}{8x} \times 100$$

$$= 25\%$$

Q.5 Rashmi buys a calculator of ₹ 720 and sells it at a loss of $6\frac{2}{3}\%$, f or how much does she sells it?

Sol. Cp of a calculator = ₹ 720

Loss percent = $6\frac{2}{3}\%$

Sp of the calculator = ?

$$Sp = Cp \left(\frac{100 - l\%}{100}\right)$$

$$= 720\left(\frac{100 - \frac{20}{3}}{100}\right)$$

$$= 720\left(\frac{300 - 20}{300}\right)$$

$$= \frac{720 X 280}{300}$$

$$= 24 x 28$$

$$= ₹ 672$$

Selling price of calculator is ₹ 672

Q. 6 Harish sold a bicycle at 8% gain, had it been sold ₹ 75 more than gain would have been 14%. Find the cost price of bicycle.

Sol. Given,

Let, the cp of bicycle = ₹ X

$$\therefore \ Sp = cp \ (\frac{100 + p\%}{100})$$

$$\Rightarrow x \ (\frac{100 + 8}{100}) + 75 = x(\frac{100 + 14}{100})$$

$$\Rightarrow \frac{108x}{100} + 75 = \frac{114x}{100}$$

$$\Rightarrow 75 = \frac{114x}{100} - \frac{108x}{100}$$

$$\Rightarrow 75 = \frac{6x}{100}$$

$$\Rightarrow x = \frac{75 \times 100}{6}$$

$$\Rightarrow x = ₹1250$$

∴Cp of bicycle is ₹ 1250

DISCOUNT

In order to increase the sales or clear the old stock the shopkeeper offers a certain percentage of rebates on the market price (MP). This rebate is known as **discount**.

Cash Discount is the discount offered by retailer to the customer or consumer at special occasion is known as cash discount. Generally it offered on list price and available when the customer buying goods on cash.

Trade Discount is the discount offered by manufactures to the retailer at a fixed rate of discount on purchasing of goods. It is offered in regular nature with fixed rate of discount.

Q. 1. The marked price of the ceiling fan is ₹ 1250. The shopkeeper allows a discount of 6% on it. Find the selling price of the fan?

Sol. Market price of fan = ₹ 1250

Discount % = 6%

Selling price = ?

$$Sp = Mp \left(\frac{100 - D\%}{100}\right)$$

$$= 1250 \left(\frac{100 - 6}{100}\right)$$

$$= 1250 \times \frac{94}{100}$$

$$\therefore Sp = ₹1175.$$

Q.2 A dealer purchase a washing machine for ₹ 7660. He allows a discount 12% on its marked price and still gain 10%. Find the market price of the machine.

Sol. Cp of washing machine = ₹7660

$$\therefore \quad Sp = cp \left(\frac{100 + p\%}{100}\right)$$

$$= 7660 \left(\frac{100 + 10}{100}\right)$$

$$= 7660 \left(\frac{110}{100}\right)$$

$$= 7660 \times 1.1$$

$$= ₹ 8426$$

$$Sp = Mp \left(\frac{100 - D\%}{100}\right)$$

$$8426 = \text{Mp} \left(\frac{100-12}{100}\right)$$

$$8426 = \text{Mp} \left(\frac{88}{100}\right)$$

$$\text{Mp} = \frac{8426 \times 100}{88}$$

$$= ₹\, 9575$$

Q.3. Find a single discount equivalent to successive discount to 20% and 10%.

Sol. Let the Mp of articles be $= 100$

After 20% discount $\quad = 100\text{-}20$

$$= \ 80$$

Again 10% discount $\quad = 10\% \text{ of } 80$

$$= \frac{10}{100} \times 80$$

$$= 8$$

Sp after two successive discount $= ₹\, 80\text{-}\, 8$

$$= ₹\, 72$$

Single rate of discount $\quad = 100\text{-}72$

$$= 28\%$$

$$(\,or\,)$$

If successive discount is applied on ₹ 100 at 20%, and 10% after the discount

The selling price will be

$$SP = MP \left(\frac{100-D\%}{100}\right)$$

$$= 100\left(\frac{100-20}{100}\right)\left(\frac{100-10}{100}\right)$$

$$= 100 \times \frac{80}{100} \times \frac{80}{100}$$

$$= 72$$

$\therefore$ Single discount rate is 100-72 = 28%

Q.4 The marked price of the TV is ₹ 18500, a dealer allows two successive discount 20% and 5%.find how much is the T.V available.

Sol. Mp of the T.V = ₹ 18500

Successive discount rate = 20% and 5%

$$Sp = Mp \left(\frac{100-D1\%}{100}\right)\left(\frac{100-D2\%}{100}\right)$$

$$= 18500\left(\frac{100-20}{100}\right)\left(\frac{100-5}{100}\right)$$

$$= 18500 \times \frac{80}{100} \times \frac{95}{100}$$

$$= ₹14060$$

Q.5 A jeweller allows a discount of ₹ 16% to its customers and still gains 20%. Find the marked price of the cost of the jewels is ₹ 1190.Also find the selling price if the discount rate is increased to 20%.

Sol. Cp of a ring = ₹ 1190

P% = 20%

$$\text{Sp} = \text{cp} \left(\frac{100+p\%}{100}\right)$$

$$= 1190\left(\frac{100+20}{100}\right)$$

$$= 1190 \times 120/100$$

$$\text{Sp} = ₹1428$$

$$\text{Sp} = \text{mp} \left(\frac{100-D\%}{100}\right)$$

$$\Rightarrow 1482 = \text{mp} \left(\frac{100-16}{100}\right)$$

$$\Rightarrow 1482 = \text{mp} \left(\frac{84}{100}\right)$$

$$\Rightarrow \text{mp} = \frac{1482 \times 100}{84}$$

$$\Rightarrow \text{mp} = ₹\,1700$$

When D% = 20, Sp=? Mp=1700

$$\text{Sp} = 1700\left(\frac{100-D\%}{100}\right)$$

$$= 1700\left(\frac{100-20}{100}\right)$$

$$= 1700\left(\frac{80}{100}\right)$$

$$= 1700 \times \frac{80}{100}$$

$$= 170 \times 8$$

$$\therefore \text{sp} = ₹1360$$

PROBLEMS FOR PRACTICE

1) The cost price of 12 candles is 3equal to the sp of 15 candles. Find loss or profit %Oranges are brought at 6 for ₹10 and sold it 4 for ₹9. Find profit or loss %.

2) A radio is sold for ₹ 3120 at a loss of 4%. What will be the gain or loss%, if it is sold for ₹3445?

3) A shopkeeper sold two fans for ₹ 990 each. One of them gains 10% and other loss 10%.Find combines gain or loss in whole transaction.

4) After allowing a discount of 8% on a toy it is sold for ₹216.20.Find marked price of the toy.

5) A man allows 10% discount on the marked price of his articles. What should be the marked price of the articles costing ₹ 600 to gain 20%.

6) A men sell an article for ₹85 at a loss of 9% in how much he should sell in order to gain $13\frac{3}{4}$ %.

Answers: 1)25%, 2)35% profit 3)6% 4)1% loss 5) ₹ 235 6) ₹ 800 7) ₹106.25

SIMPLE INTEREST AND COMPOUND INTEREST

SIMPLE INTEREST AND COMPOUND INTEREST

Simple interest: The calculation of simple interest is purposive by commercial banks, financial institution to calculate the interest payable to the investors and receivable from borrowers. It is charged on the amount originally invested (principal) at the beginning at a fixed rate is called simple interest.

$$SI = \frac{P X R X T}{100}$$

P: Principal

R: Rate of interest

T: time period

S.I: Simple Interest

$$P = \frac{SI \ X \ 100}{RT} \; ; \; T = \frac{SI \ X \ 100}{PR} \; ; \; R = \frac{SI \ X \ 100}{PT}$$

Q.1 Find the simple interest on 5000 at 10% p.a for 3 years. Also find the amount.

Sol. Given,

Principal (P) = ₹ 5000

Rate of interest (R) = 10% p.a

Time (T) =3 year

$$\therefore SI = \frac{P X R X T}{100}$$

$$= \frac{5000 X 10 X 3}{100}$$

SI $= ₹1500$

Amount = P+SI

$$= 5000+1500$$

$$= ₹6500$$

Q.2 Find the simple interest on ₹8000 at 8% p.a for 2 year 3 months.

Sol. Given

Principal (P) =₹8000

Rate of interest (R) =8% p.a

Time (T) =2 years 3 months

$$= 2\frac{3}{12}$$

$$= \frac{9}{4} \text{ years}$$

$$S.I = \frac{P X R X T}{100}$$

$$= \frac{8000 X 8 X 9}{100}$$

$$= \frac{576000}{4}$$

$$= ₹1440$$

Q. 3 P=₹15000, R=$12\frac{1}{2}$%, T=5 years 6 months, S.I =? A=?

Sol. Given;

Principal (P) =₹15000

Rate of Interest (R) =$12\frac{1}{2}$% or $\frac{25}{2}$%

Time (T) = 5 years 6 months

$$=5+\frac{6}{12}$$

$$=\frac{11}{2} \text{ years}$$

$$SI = \frac{PXRXT}{100}$$

$$=\frac{15000X25X11}{100X2X2}$$

$$=\frac{41250}{4}$$

$$=₹10312$$

Amount = P + SI

$$=15000+10312.50$$

$$=₹25312.50$$

Q.4 P=₹16000, R=$6\frac{1}{2}$%, T=4years 8 months SI=? A=?

Sol. Given;

Principal (p) = ₹16000

Rate (R) $= 6\frac{1}{2}\%$, or $\frac{13}{2}\%$

Time (T) = 4 years 8 months

$$= 4 + \frac{8}{12}$$

$$= 4 + \frac{2}{3}$$

$$= \frac{14}{3} \text{ years}$$

$$S.I = \frac{PXRXT}{100}$$

$$= \frac{16000X13X14}{100X2X3}$$

$$= \frac{29,120}{6}$$

$$= ₹4853.3$$

Q.5 Two equal sums were lent out at a 7% and 5% S.I respectively. The interest earns on the two sums up to ₹960 in 4 years. Find the sum lent in each interest.

Sol. Let, the sum lent in two different rates we notes ₹ x.

Simple interest on ₹ x @ 7% for 4 years.

$$S.I = \frac{PXRXT}{100}$$

$$= \frac{xX7X4}{100}$$

$$= \frac{7x}{25}$$

Simple interest on ₹ x @ 5% for 4 years

$$\text{S.I}=\frac{PXRXT}{100}$$

$$=\frac{xX5X4}{100}$$

$$=\frac{x}{5}$$

$$\therefore \frac{7x}{25}+\frac{x}{5}=960$$

$$\Rightarrow \frac{7x+5x}{25}=960$$

$$\Rightarrow \frac{12x}{25}=960$$

$$\Rightarrow x=\frac{960X25}{12}$$

$$\Rightarrow x=₹2000$$

$$\therefore \text{Total amount} = x+x$$

$$=2000+2000$$

$$=₹4000$$

Q.6 The S.I on certain principal on 5 years is ₹360 and the interest is $\frac{9}{23}$ part of the principal. Find the principal and the interest rate.

Sol. Given;

S.I=₹360, R=? , P=?, T=5years

$$\Rightarrow \frac{9}{25}P=₹360$$

$$P=\frac{360X25}{9}$$

P=1000

$$S.I=\frac{P \times R \times T}{100}$$

$$R=\frac{S.I \times 100}{P \times T}$$

$$=\frac{360 \times 100}{1000 \times 5}=\frac{36}{5}$$

R=7.5%

Q.7 A man invested a certain sum of money at 10% simple interest. After 1 year he invested an equal amount at 12% simple interest when the amount in each case becomes ₹1600. He withdraws the money. How much money was invested in each case after how many years the 1st one was withdrawn?

Sol. Given;

Amount (A) = ₹1600 (P+SI)

Rate (R) = 10% and 12%

Principal (P) =? (x)

Time (T) = (N) year or (n-1) year

Let, the principal invested in each case ₹ x

Time (T) = (N) year or (n-1) year

Let, the principal invested in each case ₹ x

For 1st case:

Amount =P+S.I

$$1600 = x + \frac{x + 10 X n}{100}$$

$$1600 = x + \frac{10nx}{100}$$

For IInd case:

Amount = P+S.I

$$1600 = x + \frac{x X 12 X (n-1)}{100}$$

$$1600 = x + \frac{12x(n-1)}{100}$$

Solving Equation (i)-(ii) $\Rightarrow 0 - 0 + \frac{10nx}{100} - \frac{12x(n-1)}{100}$

$$0 = \frac{10nx - 12nx + 12x}{100}$$

$$0 = \frac{-2nx + 12x}{100}$$

$$\Rightarrow -2nx + 12x = 0$$

$$\Rightarrow 12x = 2nx$$

$$\Rightarrow 12 = 2n$$

$$\Rightarrow n = \frac{12}{2}$$

$$\Rightarrow n = 6$$

From –(i)

$$\Rightarrow 1600 = x + \frac{10 X 6 X x}{100}$$

$$\Rightarrow 1600 = \frac{x}{1} + \frac{6x}{10}$$

$\Rightarrow 1600 = \frac{10x + 6x}{10}$

$\Rightarrow 1600 = \frac{16x}{10}$

$\Rightarrow x = \frac{1600 \times 10}{16}$

$\Rightarrow x = ₹1000$

Q.8 A sum of money amount to ₹20800 in 5 years and ₹22, 720 in 7 years. Find the principal and rate of interest.

Sol. Given;

P+S.I in 5 years =₹20800 ------- (i)

P+S.I is 7 years = ₹22720----- (ii)

∴ S.I in 2 years = 1920 ((ii)-(i))

∴ S.I in 1 years = $\frac{1920}{2}$ = ₹960

S.I in 5 years = 5x1960 =₹4800

P+4800=20800

P=20800-4800

P=₹16000, I=₹4800 , T=5 years, R=?

$R = \frac{S.I \times 100}{P \times T}$

$= \frac{4800 \times 100}{16000 \times 5}$

R=6%

Q.9 Ram opens a monthly deposited account in a Bank on 1.4.84 by depositing 100 at beginning of each month. The Bank offered 10% simple interest and such deposit. How much money he will get at the end of the year.

Sol. The first deposit of ₹100 will earned interest for 12 months.

The second deposit of ₹100 will earned for 11 months, similarly the last deposit of ₹100 will earned interest for one month.

∴ Total interest is the interest on ₹100 for (12+11+10+......) months

$$= \frac{12 X 13}{2}$$

$$= 78 \text{ months}$$

$$= \frac{78}{12} \ years$$

R $= 10\%$

P $= ₹100$

∴ S.I $= \frac{P X S X T}{100}$

$$= \frac{100 X 10 X 78}{100 X 12}$$

$$= ₹65$$

Amount Received at the end year.

$$= 100x12+65$$

$$= 1200+65$$

$$= ₹1265$$

Q.10 How much should a person invest in a recurring deposit every month, so that after 20 such installment he can get an amount of ₹2175 after 20 months from a bank paying 10% P. a simple interest.

Sol. Let the installment of amount deposit ₹ X. so the first installment will earn interest for 20 months.

Second installment will earned interest for 19 months similarly the last installment will earned interest for 1 month.

∴ Total interest is equal interest on ₹ X for

$$= (20+19+18+.............+1) \text{ months}$$

$$= \frac{20X21}{2}$$

$$=210 \text{ months}$$

$$=\frac{210}{12} \text{ years}$$

$$\text{S.I}=\frac{PXRXT}{100}$$

$$=\frac{x+10X210}{100X12}$$

$$=\frac{7x}{4}$$

Amount = P+I

$$\therefore 2175=20x+\frac{7x}{4}$$

$$\Rightarrow 2175=\frac{80x+7x}{4}$$

$$\Rightarrow x = \frac{2175X4}{87}$$

$$\Rightarrow x = ₹100$$

Q. 11 A person borrowed ₹500 from a money lender and agree to pay for installment of ₹150 each at the end of the 3, 6, 9 and 12 months. Find the rate of simple interest per annum charged by the moneylender.

Sol. Principal borrowed ₹500

Amount of installment paid = ₹150

No. of installment = 4

∴ Interest paid = Amount – principal

$$= 150 \times 4 - 500$$

$$= 600 - 500$$

$$= ₹100$$

$$\text{S.I} = \frac{PXRXT}{100}$$

$$\Rightarrow \frac{500XRX1}{100X4} + \frac{350XRX1}{100X4} + \frac{200XRX1}{100X4} + \frac{50XRX1}{100X4} = 100$$

$$\Rightarrow \frac{5r}{4} + \frac{7x}{8} + \frac{r}{2} + \frac{r}{8} = 100$$

$$\Rightarrow \frac{500XRX3}{100X12} + \frac{350XRX6}{100X12} + \frac{50XRX12}{100X12} = 100$$

$$\Rightarrow \frac{R}{1200}(1500 + 2100 + 1800 + 600) = 100$$

$$\Rightarrow \frac{RX6000}{1200} = 100$$

$$\Rightarrow Rx5 = 100$$

$$\Rightarrow R = \frac{100}{5} = 20\%$$

PROBLEMS FOR PRACTICE

1. Find the period in which ₹6000 at 15% simple interest would become ₹2700. (Ans: 3 years)

2. A man deposited ₹5000 in a bank on 1st Jan and end of the 6 months withdrew ₹ 3000. Find the interest due to him at the end of the year at 6% simple interest per annum.(Ans: ₹210)

3. Satish deposited a total sum of ₹5000 in two bank. One of them pays at 8% interest and other 10% at the end of 1 year he received ₹430 as interest. Find the sum he invested in two banks. (Ans: ₹3500 and ₹1500)

4. A person puts out on simple interest ₹500 for 4 years and ₹600 for 3 years and altogether receives ₹190 as interest. What is the rate per cent per annum? (Ans 5%)

5. Out of a sum of ₹1550 a part was invested at 5% annual interest and the rest at 8% simple interest. If the total interest after 3 years is ₹ 300, then what is the ratio of the amounts invested at 5% and 8%?

   (Ans 16:15)

COMPOUND INTEREST

Compound interest is more benefit to the investors comparing simple interest, because we get interest over interest accumulated with the principal amount every year. We have separate formula for finding amount which includes compound interest. Maximum of commercial banks offered compound interest to their investors and also charged compound interest to the borrowers.

There are two methods can be used to find compound Interest.

i) Simple interest formula method
ii) Compound interest formula method

The both methods has been discussed as follows,

Amount on compound (A) = P $(1+\frac{R}{100})^n$

Compound interest = A-P

$$=P \left(1+\frac{r}{100}\right) - p$$

$$C.I = P \left[\left(1+\frac{r}{100}\right)-1\right]$$

Where, R=Rate of interest, n=Time period

Q.1 Find the amount of ₹8000 for 3 year compounded annually at 5% per annum also find the compound interest.

Sol. Given,

Principal (P) =₹8000

Rate (R) = 5%

Time (N) = 3Years

Amount (A) = P $(1+\frac{r}{100})^n$

$$=8000(1+\frac{5}{100})^3$$

$$=8000(\frac{21}{20})^3$$

$$=8000 \times \frac{21X21X21}{20X20X20}$$

$$=21x21x21$$

$$A= ₹9621$$

∴CI=A-P

$$=9621-8000$$

$$=₹1261$$

Q.2 Find the compound interest on ₹6400 for two years compounded annually at $7\frac{1}{2}$ per annum.

Sol. Given,

Principal (P) = ₹6400

Rate (r) = $7\frac{1}{2}$%

Time (n) = 2 years

Amount (A) = P $(1+\frac{r}{100})^n$

$$=6400(1+\frac{15}{200})^2$$

$$=6400(\tfrac{215}{200})^2$$

$$=6400(\tfrac{43}{40})$$

$$=6400 \times \tfrac{215 X 215}{200 X 200}$$

$$=6400 \times \tfrac{43 X 43}{40 X 40}$$

$$=4 \times 43 \times 43$$

$$=₹7346$$

∴ Compound interest = A-P

$$= 7396\text{-}6400$$

$$=₹996$$

Q.3 Find the amount of ₹12000 after 2 years compounded annually the rate of interest being 5% p.a during first years and 6% p.a during second year.

Sol. Given,

Principal (p) =₹12000

Rate of interest (R_1) = 5%

$$(R_2) = 6\%$$

Time period (n) = 2 years

$$A = P \ (1+\tfrac{R1}{100})^1 \ (1+\tfrac{R2}{100})^1$$

$$=12000(1+\tfrac{5}{100}) \ (1+\tfrac{6}{100})$$

$$= 12000\left(\tfrac{21}{20}\right)\left(\tfrac{53}{50}\right)$$

$$= 12000 \times 1.05 \times 1.60$$

$A = ₹13356$

$\therefore C.I = A\text{-}P$

$$= 13356\text{-}12000$$

$$= ₹1356$$

Q.4 Find the compound interest earned ₹5000 at 10% rate of interest for $2\tfrac{1}{2}$ years.

Sol. Given

Principal (P) = ₹5000

Rate of interest (R) =10%

Time period (n) =$2\tfrac{1}{2}$ years

$A = P\left(1+\tfrac{10}{100}\right)^n$

$=5000\left(1+\tfrac{10}{100}\right)^2 \quad \left(1+\tfrac{5}{100}\right)^1$

$=5000\left(\tfrac{110}{100}\right)^2 \times \tfrac{105}{100}$

$=5000 \times \tfrac{121}{100} \times \tfrac{105}{100}$

$=5 \times \tfrac{121}{4} \times 105$

$= 5 \times 30.25 \times 105$

$=₹6352.50$

WHEN INTEREST BEING COMPOUNDED HALF YEARLY

If we invest the principal at half year basis at compound interest, we get more interest than when we invest it at annual basis.

$A = P\left(1 + \dfrac{r}{200}\right)^{2n}$

Amount, P –Principal invested, r-rate of interest, n-time periods

Compound interest = Amount - Principal

Q.1 Compute the interest on ₹1000 for 10 years as 4% per annum, the interest being paid half yearly.

Sol. Given,

P $= ₹10000$

Rate $= 4\%$ p.a

T (n) $= 10$ years

A $= ?$

A $= P\left(1 + \dfrac{r}{200}\right)^{2n}$

$= 1000\left(1 + \dfrac{4}{200}\right)^{2(10)}$

$= 1000(1.02)^{20}$

$= 1000 \times 1.486$

$= ₹1486$

$\therefore$ C.I $= A-P$

=1486-1000

=₹486

Q.2 Find the compound interest on ₹15625 for $1\frac{1}{2}$ year at 8% p.a when compounded half years

Sol. Given

P =₹15625

Time (n) =$1\frac{1}{2}$ year or $\frac{3}{2}$ years

R=8

Compound half yearly

$$A = P \left(1+\frac{r}{200}\right)^{2n}$$

$$= 15625\left(1+\frac{8}{200}\right)^{2\left(\frac{3}{2}\right)}$$

$$= 15625 \times \frac{208}{200} \times \frac{208}{200} \times \frac{208}{200}$$

$$= ₹17576$$

∴ Interest =A-P

=17576-15625

=₹1951

Q.3 Find compound interest on ₹160000 for two years at 10% p.a when compounded semi- annually.

Sol. Given,

P=₹160000

R=10%

T (n) = 2 years

Compounded half-yearly

$A = P(1+\frac{r}{200})^{2n}$

$= 160000(1+\frac{10}{200})^{2 \times 2}$

$= 160000(\frac{210}{200})^4$

$= 160000 \times \frac{210}{200} \times \frac{210}{200} \times \frac{210}{200} \times \frac{210}{200}$

$= 160000 \times 1.05 \times 1.05 \times 1.05 \times 1.05$

$= ₹194481$

$\therefore$ Interest $= A-P$

$= 194481-160000$

$= ₹34431$

Q.4 Find compound interest on ₹125000 for 9 month at 8% p.a compounded quarterly.

Sol. Given

P $= ₹125000$

T (n) $= \frac{9}{12}$ or $\frac{3}{4}$ years

R ate $= 8\%$

$A = P(1+\frac{r}{400})^{4n}$

$$= 125000 \left(1 + \frac{8}{400}\right)^{4 \times \frac{3}{4}}$$

$$= 125000 \left(\frac{408}{400}\right)^3$$

$$= 125000 \times \frac{408}{400} \times \frac{408}{400} \times \frac{408}{400}$$

$$= 125000 \times 1.02 \times 1.02 \times 1.02$$

$$= ₹132651$$

$\therefore$ Interest $= $ A-P

$$= 132651 - 125000$$

$$= ₹7651$$

PROPERTIES OF LOGARITHIMS

Logarithms are applicable by using the following properties. We can use it when the calculation is complicated to process through calculator. Many a time the solution is rather difficult. To avoid the complication we use the technique of Logarithms.

If $a^x = M$ (exponential form), then $\log_a M = x$ (logarithmic form), in this way we can reduce the exponents function into normal function.

(i) $\log(mn) = \log m + \log n$

(ii) $\log m^n = n \log m$

(iii) $\log \frac{m}{n} = \log m - \log n$

From the following information the application of logarithms may be clearer

Exponential form	Log-form $\log_a M = x$	Base
$10^2 = 1000$	$\text{Log}_{10} 1000 = 3$	10
$7^3 = 343$	$\text{Log}_7 343 = 3$	7
$2^t = M$	$\text{Log}_2 M = t$	2

Q.5 A sum of money invested at compound interest amount to ₹21632 at the end of second years and ₹22497.28 at the end of third year. Find the rate of interest and the sum invested.

Sol. Given

Let the principal invested be denote P, rate of interest be r

According to the available information

$A = P(1 + \frac{r}{100})^n$

After 2 year

$\Rightarrow 21632 = P(1 + \frac{r}{100})^2$ ------------ (I)

After 3 years

$\Rightarrow 22497.28 = P(1 + \frac{r}{100})^3$ ------------ (II)

Divide (ii) by (i)

$\Rightarrow \dfrac{22497.28}{21632} = \dfrac{P(1 + \frac{r}{100})^3}{P(1 + \frac{r}{100})^2}$

$\Rightarrow 1.04 = 1 + \dfrac{r}{100}$

$\Rightarrow 1.04 - 1 = \dfrac{r}{100}$

$\Rightarrow 0.04 = \dfrac{r}{100}$

R=4%

From (1) equation

$\Rightarrow 21632 = P\left(1+\dfrac{4}{100}\right)^2$

$\Rightarrow 21632 = P\,(1.04)^2$

$\Rightarrow \dfrac{21632}{(1.04)^2} = P$

$\Rightarrow P = \dfrac{21632}{1.0812}$

$\Rightarrow P = ₹20000$

∴ The rate of interest is 4% and the sum invested is ₹20000

Q.6 The population of a country increase every by 2.4% of the population at the beginning of the year. In what time will be population doubles itself? Answer to the nearest year.

Sol. Let the initial population be x and after some years it become 2 x at 2.4% rate.

By using amount formula;

$\Rightarrow 2x = x\left(1+\dfrac{2.4}{100}\right)^n$

$\Rightarrow \dfrac{2x}{x} = \left(1+\dfrac{2.4}{100}\right)^n$

$\Rightarrow 2 = (1+0.074)^n$

$\Rightarrow 2 = (1.074)^n$

Applying log.

$\text{Log}_2 = \log (1.024)^n$

$\text{Log}_2 = n \log (1.024)$

$N = \dfrac{\log 2}{\log 1.024}$

$= \dfrac{0.3010}{0.0103}$

$= 29.2233$

X=29 years (approx)

Q.7 A sum of ₹1200 become ₹1322 is 2 years at compound interest compounded annually. Find the rate percent.

Sol. Given, P=1200 A=₹1322, T=2 years, R=?

$A = P \left(1+\dfrac{r}{100}\right)^n$

$\Rightarrow 1322 = 1200\left(1+\dfrac{r}{100}\right)^2$

$\Rightarrow \dfrac{1322}{1200} = \left(1+\dfrac{r}{100}\right)^2$

$\Rightarrow 1.1017 = \left(1+\dfrac{r}{100}\right)^2$

$\Rightarrow \log (1.1017) = \log \left(1+\dfrac{r}{100}\right)^2 = 2 \log \left(1+\dfrac{r}{100}\right)$

$\Rightarrow \dfrac{1}{2}\log (1.1017) = \log \left(1+\dfrac{r}{100}\right)$

$\Rightarrow \dfrac{1}{2}(0.0416) = \log \left(1+\dfrac{r}{100}\right)$

$\Rightarrow 0.0216 = \log \left(1+\dfrac{r}{100}\right)$

$\Rightarrow$ Antilog $(0.0216) = (1+\frac{r}{100})$

$\Rightarrow 0.1051 \times 10 = 1+\frac{r}{100}$

$\Rightarrow 1.051 = 1+\frac{r}{100}$

$\Rightarrow 1.051 - 1 = \frac{r}{100}$

$\Rightarrow 0.051 = \frac{r}{100}$

$\Rightarrow r = 5.1\%$

$\Rightarrow r = 5\%$ (approx)

Q.8 On what sum of many will the different between the simple interest and the compound interest for 2 year at 5% per annum be equal to ₹50?

Sol. Given,

P=?

T=2years

R=5%

C.I-S.I=₹50

$\Rightarrow P\left[(1+\frac{r}{100})^n - 1\right] - \frac{P \times R \times T}{100} = 50$

$\Rightarrow P\left[(1+\frac{r}{100})^2 - 1\right] - \frac{P \times 5 \times 2}{100} = 50$

$\Rightarrow P\left[(1.05)^2 - 1\right] - 0.IP = 50$

$\Rightarrow P\left[0.1025\right] - 0.IP = 50$

$\Rightarrow$P (0.0025) = 50

$\Rightarrow$P= $\frac{50}{0.0025}$

$\Rightarrow$P=₹20000

Q.9 Machine deprecation at the rate of 7% of its value at the beginning of a year. If the machine was purchase for ₹ 8500 what is the minimum number of complete years at the end of which the worth of the machine will be less than or equal to half of its original cost price?

Sol. Given,

Value of machine (P) =₹8500

Rate of Interest (r) =7%

After some years (P^1) =4250

$P^1 = P (1 - \frac{r}{100})^n$

$4250=8500 (1-\frac{7}{100})^n$

$=\frac{1}{2}= (1-0.07)^n$

$=\frac{1}{2}= (0.93)^n$

$\text{Log} \frac{1}{2}= \log (0.93)^n$

$\text{Log} (0.5) =n \log (0.93)$

$n=\frac{\log(0.5)}{\log(0.93)}$

$$n=\frac{-1+0.6990}{-1+0.9685} \quad \text{(using log table)}$$

$$n=\frac{-0.301}{-0.315}$$

$$n=9.55$$

$$n=10 \text{ years (approx)}$$

Q.10 Find the compound interest by the simple interest formula? P=₹10000, R=10% T=3years C.I=?

Sol. Given,

P=₹10000

R=10%

T=3 years

C.I=?

Simple interest for 1st years:

$$S.I=\frac{P \times R \times T}{100}$$

$$=\frac{10000 \times 10 \times 1}{100}$$

$$=₹1000$$

S.I for 2nd years:

$$S.I=\frac{P \times R \times T}{100}$$

$$=\frac{(10000+1000) \times 10 \times 1}{100}$$

$$=\frac{11000 \times 10 \times 1}{100}$$

=₹1100

S.I for 3rd years

$S.I=\frac{PXRXT}{100}$

$=\frac{(1100+11000)X10X1}{100}$

$=\frac{12100X10X1}{100}$

=₹1210

∴ C.I=1000+1100+1210

=₹3310

Checking

$A=P\ (1+\frac{r}{100})^{3}$

$=10000(1+\frac{10}{100})^{3}$

$=\frac{10000X1331}{1000}$

A=₹13310

∴ C.I=A-P

=13310-10000

=₹3310

Q.11 P=₹50000, R=8%, T=3years, C.I=?

Sol. Given

P=₹50000

R=8%

T=3years

C.I=?

Simple interest for 1st years

$$S.I = \frac{PXRXT}{100}$$

$$= \frac{50000X8X1}{100}$$

$$= ₹4000$$

Simple interest for 2nd years

$$S.I = \frac{PXRXT}{100}$$

$$= \frac{(50000+4000)X8X1}{100}$$

$$= \frac{54000X8}{100}$$

$$= ₹4320$$

Simple interest for 3rd years

$$= \frac{(54000+4320)X8X1}{100}$$

$$= \frac{58X320X8}{100}$$

$$= 4665.60$$

∴ C.I = 1st + 2nd + 3rd years S.I

=4000+4320+4665.60

=₹12985.60

∴Compound interest for 3 the years is ₹12985.60

Practice exercise

1) Principal ₹ 12000, Rate 6%, Time period 2 years. Find compound interest. **Ans: ₹1483.20**.

2) Principal ₹10000, Rate 12%, Time $1\frac{1}{2}$ years. Find compound interest. **Ans: `1872**

3) Principal ₹25000, Rate $7\frac{1}{2}$ %, Time $2\frac{1}{2}$ years. Find compound interest. **Ans: ₹ 4974**

4) Swathi borrowed ₹ 40960 from a bank to buy a piece of land if the bank charges $12\frac{1}{2}$ %P.a compounded half yearly. What amount will she have to pay after $1\frac{1}{2}$, if the interest is compounded annually? What is excess amount she has to pay to the bank? **Ans: ₹210**

5) Compute the interest on ₹ 5000 for 10 years at 4% P.a, if the interest being compounded annually. **(Ans: ₹2401.20)**

6) Compute the interest on ₹10000 for 7 years at 6% P.a the interest being paid annually. **Ans: ₹5036.30**

7) Find the compound interest ₹5000 for 3 years if the rate of interest being 5%, 6%and 7% for the 1st, 2nd and 3rd years respectively. **Ans: ₹ 954.55**

SHARES DEBENTURES AND STOCKS

DIVIDEND

Dividend is a part of profit of company which is distributed to the shareholders as per the numbers of share purchase by them or fixed percentage on the share capital (face value of share)

FACE VALUE

The value for which a share is issued originally by a company at the time of fresh issue is called the face value of the share. It is printed on the share certificate and also knows as normal or par value of the share.

MARKET VALUE

The value for which a share is available in the market is called the market value of the share; this may by vary from day to day where its face value always remains the same. The market value of the share is above the par is called premium and below the par is called discount.

BROKERAGE

The sale or purchase of share is generally done through agents called share broker or broker. They charge a minimum amount from buyer as well as sellers for their services. This charge is called brokerage. It is charged on market value of the share.

COMMISSION

The commission is the amount earned by a manager or agent on their target achieved during a specific period. It will be varied from business to business according to the nature and amount of transaction. It is charged on market value of shares.

Q.1 Company issued share at 10% premium. Satish applied 1000 share but he allotted only 500 shares. Find the investment if the face value of share is ₹100.

Sol.

Given, Number of applied shares = 1000

Number of allotted shares = 500

Face value =₹100

Premium =10%

Market value = ₹100+10

$$=₹110$$

∴ Investment by Satish = 500 x 110 =₹55000

Q.2 Find the investment in buying 525 shares of ₹100 each at ₹12 premium.

Sol. Given, Number of share = 525

Face value = ₹100

Premium =₹12

Market value = 100+12

=₹112

∴ Investment = 525x112

=₹58800

Q.3 Find the investment in buying 450 shares at ₹100 each. At 5 % discount.

Sol. Given, Number of shares = 450

Face value = ₹100

Discount = ₹5

Market value = 100-5

=₹ 95

∴ Investment = 450x95

=₹42750

Q.4. A company issued 50000 shares of par value of ₹10 each. If the total dividend declared by the company is ₹62500 find the rate of dividend paid by the company?

Sol.

Given, No. of share issued = 50000

Par value of a share = ₹10

Amount of dividend declared = ₹62500

Rate of Divided =?

Share capital = 50000x10

$$=₹5000000$$

Rate of Dividend $= \frac{62500}{500000}$x100

$$=12.5\%$$

Q.6. Ravi bought 500 shares of a company quoted at ₹280. Find the amount spends by him on this purchase.If the brokerage be 1%, find the cost of buying the shares.

Sol. Given, No. of share purchase = 500

Market value for share = ₹280, Brokerage = 1%

Total amount spend on the purchase of 500 share

= Market value of the share + brokerage (1%)

=500x280+1%of (140000)

=140000+1400

=₹141400

Q.7. Shalu had 50 preference share and 400 common share of par value ₹100 each. If the dividend declares on preference share is 10% p.a and semi – annual dividend of 7.5% is on common share. Find the annual dividend received by shalu.

Sol.

Given, No. of preference share = 50

Par value = 100 per share

Rate of dividend $= 10\%$

No. of common share $= 400$

Par value $= ₹100$ p/s

Rate of dividend $= 7.5\%$ half yearly

$\qquad = 15\%$ p.a

Dividend on preference share $= 10\%$ (50x100)

$\qquad\qquad\qquad\qquad = 10\%$ of 5000

$\qquad\qquad\qquad\qquad = ₹500$

Dividend on common share $= 15\%$ of (400x100)

$\qquad\qquad\qquad\qquad = 15\%$ of 40000

$\qquad\qquad\qquad\qquad = ₹6000$

$\therefore$ Total dividend Received $= 500+6000$

$\qquad\qquad\qquad\qquad = ₹6500$

Q.8. Ankhus invested ₹4444in the shares of face value of ₹100 each of a company at the end of the year the company declare dividend at 15%. Which give him an annual income of ₹600? At what price was the share quoted if the brokerage of 1%.

Sol.

Given, No. of investment $= ₹\ 4444$

Face value $= ₹100$ each

Rate of dividend $= 15\%$

No. of annual income = ₹600 (dividend)

Brokerage = 1%

No. of share =?

Let, the No. of share will be X

∴ Par value of all share = 100x

Here, 15% of (100x) =600

$=\frac{15}{100}x100x = 600$

$X=\frac{600}{15}$

X=40 shares.

Let, the quoted price share be y

Market value of y share + 1% brokerage: 4444

40y + 1% of 40y = 4444

$40y+\frac{40y}{100} = 4444$

$=\frac{4000y+40y}{100} = 4444$

$=\frac{4040y}{100}=4444$

$Y=\frac{4444X100}{4040}$

Y=₹110

Checking

40x110+1% of 4400

=4400+44

=4444

Q.9. Renu buys 200 shares each of pas value ₹ 10 of a company which pays annual dividend of 15%. At such a price that he get 12% investments and the value of the share.

Given,

Number of share = x

Par value = ₹ 10

Rate of dividend = 15%

∴ Dividend = 15% of 15%

$$=\frac{15}{100}x10 =1.50$$

Let, the market price of the share = ₹ y

∴ $\frac{12}{100}$ X y =1.50

$$Y=\frac{1.50 X 100}{12}$$

Y= ₹ 12.50

∴ Market value of the share ₹ 12.50/-

Q.10. The capital of a company is made up of 50000 preferred shares and 20000 common shares with dividend of 20% on preferred shares. The par value of each types of being ₹10. The company has total profit of ₹ 180000 out of which ₹ 30000 were kept in reserveand the remaining distribution to shareholders. Find the dividend percent paid to the common shareholders.

Sol. Given

Number of share = 50000

Face value = ₹ 10 per share

∴ Total face value = 50000 x 10

$$= ₹ 500000$$

Dividend on the share = 20% of 500000

$$= ₹ 100000$$

Total profit = ₹ 180000

Provision for reserve = ₹30000

Remaining profit after reserve and preference dividend

=180000-30000-100000

=₹ 50000

Face value of 20000 common shares

=20000x10

=₹200000

% of dividend $=\frac{50000}{200000}$x100

=25%

Q.11. Mohit invested ₹3333 in share of face value of ₹ 100 each of a company at the end of the year the company declared 15% dividend and he get income of ₹450. At what share of price was the share quoted the brokerage being 1%.

Sol. Given

No. of invested = ₹3333

Face value =₹100

Rate of dividend = 15%

Brokerage = 1%

Income = ₹450

Let the no. of share will be x

∴ Face value of all share = 100x

Here, 15% of (100x) = 450

$\frac{15}{100}$X100x = 450

$X = \frac{450}{15}$

X=30

Let, the quoted price per share be y

Market value of y share +1% of brokerage = 3333

30y + 1% of 30y = 3333

30y + $\frac{30y}{100}$ = 3333

=$\frac{3000y+30y}{100}$=3333

$$=\frac{3030y}{100} = 3333$$

$$Y = \frac{3333 \times 100}{3030}$$

$$Y = \frac{333300}{3030}$$

$$Y = 1100$$

Q 12. The company with 70000 share par value ₹100 each. The director announced an annual dividend of 5%. Find the total annual dividend paid by the company and what is the amount of dividend will Ashok received annually if he owns 100 shares.

Sol.

Given No. of share = 70000 share

Par value= ₹100 each

% total dividend = 5% of (70000x100)

=5% of 7000000

= ₹350000

No. of share owned by Ashok = 100

∴ Total Dividend receive by Ashok

=5% of (100x100)

=5% of 10000

=₹500

Q 13. Mohan sell 5000 common share (par value ₹10) of a company A which pays a dividend of 20% at ₹30 per share. He invest the sales processed in ordinary share (each par value ₹25), of a company that pays of dividend 15%. If the market value of the share of company B is ₹40 find.

i) The no. or share of co. B purchase by Mohan

ii) The change in dividend income of Mohan.

Sol. Amount received by Mohan on selling 5000 shares of company A₹30 per share

= 5000 x 30 =₹ 150000

Market value of company B = ₹ 40

No. of share purchased = $\frac{150000}{40}$

$\qquad$ = ₹ 3750 share

Par value per share of company B = ₹ 35

∴ Total investment = 3750 x 25

$\qquad$ = ₹ 93750

% Dividend by company B = ₹ 15

∴ Dividend Received = 15% of 93750

= ₹14062.50

% Dividend by company A = 20%

∴ Dividend Received = 20% of (5000x10)

$\qquad$ =20% of 500000

$\qquad$ =₹10000/-

Q.14. Shikha bought 800 shares (pre) of par value ₹50 each at ₹2 premium (Brokerage 5%). After a month. The price of share fallen and she sold 300 of the share at ₹47 each (Brokerage 2%) and on the same date she sold the remaining share of ₹54 each (Brokerage 0.50% per share). Find gain or loss in the transaction.

Sol.

Given, No. of share bought by Shika = 800

Per value = ₹50 per share

Premium = ₹2 per share

Brokerage = 5%

Market value of 800 share = 50+2

∴ Market value of 800 share = 800x52

$$= 41600$$

Brokerage = 5% of 41600

$$= ₹ 2080$$

Total amount spent on buying 800 shares

$$= 41600 + 2080$$

$$= ₹43680 \ (cp)$$

Value of 300 share sold by Shika = 300x47

$$= ₹14100$$

Brokerage paid = 2% of 14100

$$= ₹282$$

Net value of 300 share= 14100-282=₹13818

Value of Receiving 500 share sold by Shika

$$=500 \times 54$$

$$=₹27000$$

Brokerage paid = 500x54 =₹250

∴ Net value of 500 share = 27000-250

$$=₹26750$$

∴ Total sale consideration of 800 shares

$$=13818+26750$$

$$=₹40568 \text{ (sp)}$$

Loss =43680-40568

$$=₹3112$$

Q.15. What rate % will a man get from his 200 common share of par value of ₹25 each bought at ₹5 premium, the rate of dividend being 8%.

Sol. Given, No. of common share =200

Par value =₹25

Premium = ₹5

Rate of Dividend = 8%

Face value of 200 share = 200x25

$$=5000$$

Market value of 200 share $=200 \times 30$

$$=6000$$

Dividend received $= 8\%$ of 5000

$$=₹400$$

Rate of return on his investment $=\dfrac{4000}{6000} \times 100$

$$=\dfrac{400}{6} = 6.67\%$$

STOCKS

The stock of a company is the total value of paid up shares. The **stock** of a company is sold in units called **shares**. A share is a unit of ownership, or equity, in a company or a corporation. Shares are one of the most traded financial instruments.

If you buy a share of a company, you are buying a piece of the company. When you own more than one share in a company or several companies, these are called stocks, because "stock" generally refers to a portfolio of shares.A:

For example, "stock" is a general term used to describe the ownership certificates of any company, in general, and "shares" refers to a the ownership certificates of a particular company. So, if investors say they own stocks, they are generally referring to their overall ownership in one or more companies.

Q.16 How much 6% stock at ₹ $108\frac{9}{10}$ can be ₹14300 Brokerage $1\frac{1}{10}$?

Sol. Cost of ₹ 100 share $= 108\frac{9}{10} + 1\frac{1}{10}$

$$= 108.9 + 1.10 = ₹ 110$$

Value of stock purchase $= 14300 \text{x} \frac{100}{110}$

$$= ₹ 13000$$

Q.17 Find the cost of ₹7000 of 15% stock at ₹105.

Sol. Cost of ₹100 stock = ₹105

Cost of ₹7000 stock $= 7000 \text{x} \frac{105}{100}$

$$= ₹7300/-$$

Q.18. Find the investment required to purchase ₹125000 of 8% stock at 92.

Sol. Cost of ₹100 stock is available in ₹92

∴ Cost of ₹ 125000 stock $= 125000 \text{x} \frac{92}{100}$

$$= ₹115000/-$$

Q.19 Find the cost of ₹15000 of $5\frac{1}{2}$ % stock at 99. Brokerage being Re1.

Sol. Cost of ₹100 stock is available at ₹99

Brokerage = ₹ 1

∴ Cost = 99+1= ₹ 100

Cost of ₹15000 stock $= 15000 \text{x} \frac{100}{100}$

$$= ₹ 15000/-$$

Q.20. How much 11% stock at 97 can be brought by investing ₹24250.

Sol. Original value of ₹97 stock = 100

∴ Original value of ₹ 24250= $24250 \times \frac{100}{97}$

Value of stock = ₹25000/-

Q.21 What amount of stock can be brought 1133 in 4% stock at $102\frac{9}{10}$ brokerage $\frac{1}{10}$?

Sol: Market value of ₹100 stock $= 102\frac{9}{10} + \frac{1}{10}$

$$= \frac{1029}{10} + \frac{1}{10}$$

$$= \frac{1030}{10}$$

$$= ₹103$$

Amount of investment = ₹1133

∴ Value of stock for ₹1133= $1133 \times \frac{100}{103}$

$$= ₹1100/-$$

Q.22. How much 7% stock at $106\frac{1}{10}$ is sold to realize ₹ 21000? Brokerage being $1\frac{1}{10}$?

Sol. Market value of ₹ 100 stock $= 106\frac{1}{10} - \frac{1}{10}$

$$= \frac{1061}{10} - \frac{1}{10}$$

$$= \frac{1061 - 11}{10}$$

$$= ₹105$$

Amount of investment = ₹21000

∴ Value of stock for ₹21000= $21000 \times \frac{100}{105}$

$$=₹20000$$

Q.23 How much 4% sold at $84\frac{1}{4}$ to realized ₹12600, brought being $\frac{1}{4}$?

Sol. Market value of ₹100 stock = $84\frac{1}{4} - \frac{1}{4}$

$=\frac{337-1}{4}$

$=\frac{336}{4}$ =₹84

Amount of investment =₹ 12600

∴ Value of stock for ₹12600=$12600 \times \frac{100}{84}$

$$=₹15000$$

Q.24. Ravi sells ₹7337 stock at 6% stock at 96 invested 8% stock at $106\frac{2}{3}$. How much stock does he hold now?

Sol. Market value of ₹100 stock = ₹96

∴ Market value of ₹ 7337 = $7337 \times \frac{96}{100}$

$$= ₹ 7043.52$$

Cost of ₹ 100 stock = $106\frac{2}{3} = \frac{320}{3}$

Cost of ₹7043.52 $\times \frac{100}{320}$

$$= \frac{7043.52 \times 300}{320}$$

$$= ₹6603.30/-$$

Q.25. How much money sustains by selling ₹30000 stock at 93 Brokerage may ₹ $1\frac{1}{2}$?

Sol. Market value of ₹100 stock = $93-1\frac{1}{2}=93-1.5$ $=91.5$

∴ Amount realize for selling ₹30000 share $=30000 \times \frac{91.5}{100}$

$=₹27450$

Q.26. Find the change in income by transferring ₹ 6000 of stock at 90 to a $1\frac{1}{2}$% of stock at

Sol. Income on ₹ 6000 stock at $13\frac{1}{2}$%

$$= 13\frac{1}{2}\% \text{ of } 6000$$

$$= \frac{27}{200} \times 6000$$

$$= ₹810/-$$

Amount realized on ₹ 6000 stock at 90 par.

$$= 6000 \times \frac{90}{100}$$

$$= ₹5400$$

∴ Income on ₹5400 stock at $10\frac{1}{2}$%

$$= 10\frac{1}{2}\% \text{ of } 5400$$

$$= \frac{21}{200} \times 5400$$

$$= ₹\, 567/\text{-}$$

Difference in income Receive

$$= 810\text{-}567$$

$$= ₹243/\text{-}$$

Debentures

Debentures are securities issued to the public by any company to raise the capital fund to diverse the business. Sometimes a company needs money for further expansion and diversification of its activities. For this, the company may not issue new shares but many borrow this money from the public through issue of debentures. So, it is documentary security of acknowledgement of debt borrowed from the public and fixed % shows the interest paid by the company on the face value up to the redemption of the company.

Q.27 Compute the annual yield percent on 12% debentures of face value ₹ 100 each and available at₹ 80 each.

Sol. It is given that,

Face value of debenture $= ₹100$

Market value of a debenture$= ₹ 80$

Rate of interest $= 12\%$

Now, by investing Rs 80, annual interest = ₹ 12

∴ By investing Rs 100, annual interest = ₹ $(\frac{12}{80}x100)$=Rs 15

Hence, annual yield=15%

Q.28 Find the percent income of buyer on 6% debentures of face value. ₹100 available in the market for ₹ 150.

Sol. We have,

Market value of a debenture =Rs150

∴ Income on Rs 150= ₹ 6

⇒ Income on Rs 100 =₹ $(\frac{6}{150}x100)$ =₹ 4

Thus, there is 4% income on the debentures.

Q.29 **which** is better investment: 15% debentures at 8% premium or 14% debentures at 4% discount?

Sol. Case: I, If face value of debenture is₹ 100, Market value =₹ 108

Annual interest earned =₹ 15

∴ Annual yield =₹ $(\frac{15}{108}x100)$ = ₹13.88

Case: II, if face value of debenture is₹100, Market value = ₹ 96

Annual interest = ₹14

∴ Annual yield = ₹ $(\frac{14}{96}x100)$ =₹ 14.58

In **case** II yield more income than **Case** I. So, second alternative is the better investment.

Q.30 SomDutt invested a certain sum in 18% debentures of the face value of ₹ 100 each available at ₹90 and earned at annual income of ₹ 8100. Find the amount invested by him, if the brokerage is 1%

Sol. To purchase one debenture

Amount invested =90+1%of90=₹ 90.90

Income received as a interest = ₹ 18

So, annual income yield by investing `90.90 = ₹ 18

∴ Amount invested to earn income of₹ 8100 =₹ $(\frac{90.90}{18}$x8100)

=Rs 40905.

Hence, the amount invested by SomDutt =₹ 4090

PROBLEMS FOR PRACTICE

Q.1 Find the investment in buying 400 shares of ₹10 each at 2% discount.

Q.2 Find the investment in buying 600 shares of ₹ 10 each quoted at ₹ 18.

Q.3 Find what a buyer would have to pay for 800 shares of ₹ 10 each quoted at ₹65. What would be the gain to the share-holder, if he had purchased the share at par?

Q.4 A company declared a semi-annual dividend of 7.5%.Find the annual dividend of Rohit, owning

2250 shares of the company having a par value of ₹10 each.

Q.5 How many shares of market value ₹12.50 each can be purchased for ₹12625, if the brokerage paid is 1%?

Q.6 Find the cost price of ₹2400 of 3% stock at 20% discount.

Q.7 What amount of stock can be bought by investing ₹1560 in 8% stock at ₹104?

Q.8 A man bought ₹5000 stock at 95 and sold it when its price raisesto $98\frac{1}{2}$. Find his gain.

Q.9 Find dividend obtained from ₹5300 if 8% stock.

Q.10 A person invests ₹ 7000 partly in 3% stock at 96 and partly in 4% stock at 120 if his total income be $3\frac{4}{14}$ % on his investment, how much did he invest in each?

Q.11 How much money has a person invested in $10\frac{1}{2}$ % stock at 90 when his income is ₹ 630?

Q.12 Find the annual yield % on 16% debentures of face value Rs100 each and available at Rs 80 each.

Q.13 which is better investment: 10% debentures at 10% discount or 20% debentures at 10% premium?

Q.14 What amount of money does Rashmi get on selling 14% debentures worth Rs 16000 at

10% discount, the face value of each being Rs100 and brokerage 1.5%?

Q.15 Anish has 1200 shares of par value of Rs20 each of a company and 500 debentures of par value Rs100 each of the company. The company pays an annual dividend of 8% on the share and an interest of 12% on debentures. Find the total annual income of Anish and rate of return on his investment.

FUNCTIONS AND RELATIONS

Meaning of relation

A relation from a set 'A' to a set' B 'can be defined as R: $\{(a, b) \in A \times B\}$ if two objects or quantities are related if there is a recognizable connection or link between the two objects or quantities.

For example

$\{(a, b) \in A \times B : a \ is \ brother \ of \ b\}$

$\{(a, b) \in A \times B : a \ is \ sister \ of \ b\}$

$\{(a, b) \in A \times B : age \ of \ a \ is \ greater \ than \ age \ of \ b\}$

Types of Relation

Empty relation: A relation R in a set A is called empty relation, if no element of A is related to any element of A, i.e., $R = \Phi \subset A \times A$.

Universal Relation: A relation R in a set A is called universal relation, if each element of A is related to every element of A, i.e., $R = A \times A$

Both the empty relation and the universal relation are sometimes called **trivial relations**.

Reflexive Relation: A relation R is said to be reflexive, if aRa is exist for all $a \in A \ or \ R \ is \ exist \ (a, a) \in R$

Symmetric Relation: A relation R is said to be symmetric if there exist aRb $\Rightarrow bRa$ or $(a, b) \in R \Rightarrow (b, a)$

Transitive Relation: A relation R is said to be transitive if R satisfy that aRb, bRc $\Rightarrow aRc$ or $(a, b), (b, c) \in R \Rightarrow (a, c) \in R$

Equivalence Relation: A relation R is said to be equivalence relation if R is reflexive, symmetric and transitive.

Q.1 If R is the relation in N={1,2,3, ...} R={$(a, b): a \leq b$}. Check whether R is reflexive, symmetric or transitive.

Sol. Reflexive: Every number is equal to it, i.e $1 \leq 1, 10 \leq 10$ etc

It means a$\leq a \Rightarrow (a, a) \in R$.so it reflexive.

Symmetric: For every natural number if a$\leq b$ not $means$ $that$ $b \leq a$

i.e. (a, b) $\in R \Rightarrow (b, a) \notin R. So$ it not $symmyetric$

Transitive: For every natural number a≤b, b≤c, we have a≤c

i.e. (a, b) $\in R$, (b, c) $\in R \Rightarrow$ (a,c) $\in$R.So it transitive.

We verified the relation that it is reflexive and transitive but not symmetric so it is not equivalence relation.

Q.2. Let R is the relation of lines in the set of parallel lines in the plane. Determine whether R is reflexive, symmetric and transitive and so it is equivalence.

Sol. We can define the relation $R:\{(L_1, L_2): L_1 \parallel L_2 \text{ for all lines in the set } S\}$

Reflexive: Let L∈S, so L is parallel to L, i.e. (a, a) ∈R.So, it is reflexive.

Symmetric: LetL_1, $L_2 \in S$ and $(L_1, L_2) \in R \Rightarrow L_1 \parallel L_2$

$$\Rightarrow L_2 \parallel L_1$$

$$\Rightarrow (L_2, L_1) \in R$$

i.e. (a, b) ∈R⇒ (b, a) ∈R.So, it symmetric.

Transitive: Let L_1, L_2, $L_3 \in S$ and $(L_1, L_2) \in R$ and $(L_2, L_3) \in R$

$$\Rightarrow L_1 \parallel L_2 \dots\dots\dots (i)$$

$$\Rightarrow L_2 \parallel L_3 \dots\dots\dots (ii)$$

From (i) and (ii) $\Rightarrow L_1 \parallel L_3$

i.e (a, b) and (b, c) ∈R⇒ (a, c) ∈R

Here R is reflexive, symmetric and transitive so R is equivalence relation.

Q.3. Show that a relation R in the set Z of integers given by $R=\{(a, b): 2 \text{ is divided } (a - b)\}$ is an equivalence relation.

Sol. Reflexive: Let a∈Z, and a – a =0 is divisible by 2

i.e (a, a) ∈ R.So, it is reflexive

Symmetric: Let a, b $\in$Z and (a, b) $\in$R $\Rightarrow$ a – b is divisible by 2

$\Rightarrow$ - (a – b) is divisible by 2

$\Rightarrow$ b- a is divisible by 2

$\Rightarrow$ (b, a) $\in$R

i.e (a, b) $\in$R$\Rightarrow$ (b, a) $\in$R .So, it is symmetric.

Transitive: Let a, b, c $\in$Z and (a, b), (b, c) $\in$R

$\Rightarrow a - b$ is divisible by 2......... (i) and

$\Rightarrow b - c$ is divisible by 2......... (ii)

Adding (i) and (ii) $\Rightarrow a - c$ is divisible by 2

$\Rightarrow (a, c)$ $\in$R

So it is transitive.

Here R is reflexive, symmetric and transitive and hence it is an equivalence relation.

PROBLEMS FOR PRACTICE

Q.1 Show that the relation R in the set$\{1,2,3\}$ given by R $=\{(1,2),(2,1)\}$ is symmetric but neither reflexive nor transitive.

Q.2 Show that the relation R in the set A of all the books in a library of a college, given by R=$\{(x,y): x$ and y have same number of pages$\}$ is an equivalence relation.

Q.3 Check whether the relation R in R defined by R $=\{(a,b): a \leq b^3\}$ is reflexive, symmetric or transitive.

Q.4 Find the relation defined in set
A = {1,2,3, ... .13,14} defined as
R= $\{(x,y): 3x - y = 0\}$

Determine whether the above relation is an equivalence relation.

Functions: As the concept of function is of paramount importance in mathematics and among other disciplines as well, we would like to extend out in our business field.

Definition: Let A and B be two non empty sets. Suppose to each element of set A is assignees a unique element of the set B by a rules f then f is called a function from A to B and is denoted by f: A→**B**.

TYPES OF FUNCTIONS:

Meaning of Function:

The concept of function is of paramount importance in Mathematics .The relation between two sets can be defined as functions if there exist for every pre-image have atleast one image in the relation.

Definition of Function:

Let A and B be two non-empty sets. A relation f from A to B, i.e., a subset of AXB is called a function from A to B if

i) for each a∈A there exists b∈B such that
(a,b) ∈f

ii) (a,b) ∈f and (a,c) ∈f ⇒b=c

1. INTO FUNCTION: A function f from A to B is called a into function if its image is a proper subset of B

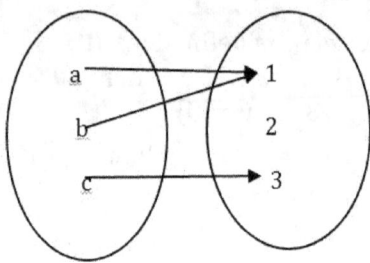

Here the above array diagram show into function.

2. ONTO FUNTION: A function f from A to B is called an onto function if its range is equal to the co-domain (i.e.) for every element in B there is a pre image in A.

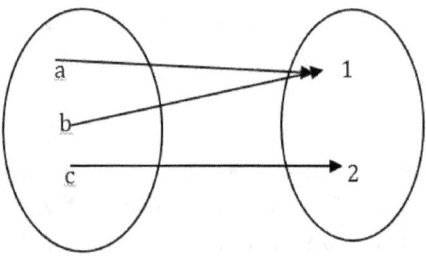

Here the above array diagram show the onto function.

3. ONE-TO-ONE FUNCTION: A function from A to B is called as one to one function if distinct elements of A are mapped into distinct elements of B.

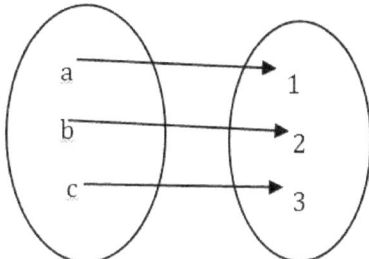

If one to one function the range is equal to the co-domain then the function is called a one to one function.

4. MANY TO ONE FUNCTION: A function from A to B is called many to one function if more than one element of A is mapped into the same image in B

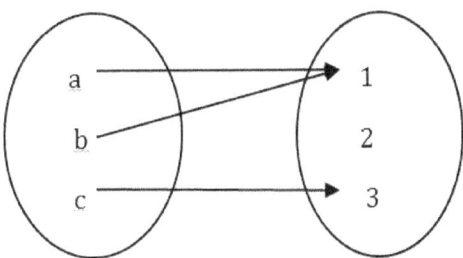

If the range is also equal to the co-domain then the function is called many to one into function.

5. CONSTANT FUNCTION: A function f from A to B is called a constant function if every element of A has the same image in B.

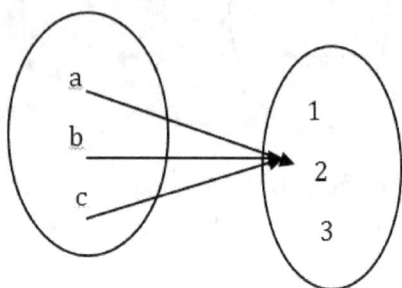

6. IDENTITY FUNTCTION: A function f from A to B is called an identify function if the image of every element is A is itself. For e.g. f(x) = x

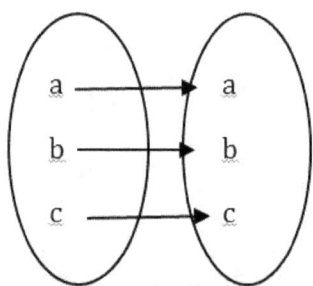

7. EVEN FUNCTION: A function f is said to be an even function if f (-x) = f(x), for all x∈R is called even function. For e.g. $f(x) = x^2$, $f(x) = x^4$ etc for any value of x f(x) is positive.

$(-1)^2 = 1$, $(-2)^2 = 4$, $(-1)^4 = 1$, $(-2)^4 = 16$

8. ODD FUNCTION: A function f is said to be odd function if f (-x) = - f(x), for some x∈R is called odd function. For e.g. F(x) = x^3, f(x) = x^5 etc. for any negative value of x is negative.

$(-1)^3 = -1$, $(-2)^3 = -8$, $(-1)^5 = -1$

9. MONOTONE FUNCTION: A function f is called monotone, if it either increasing or decreasing for any x∈R and f is called strictly monotone, if it is either strict increasing or either strict decreasing.

For e.g. Let $x_1 > x_2 \Rightarrow f(x_1) > f(x_2)$ i.e. f(x) = x-1

   If $x_1 < x_2 \Rightarrow f(x_1) < f(x_2)$ i.e. (x) = x+1

10. POLYNOMIAL FUNCTION: A function f is defined by f(X) = $a_0 + a_1 x + a_2 x^2 + a_3 x^3 + ... + a_n x^n$ where a_0, a_1 a_3 ......a_n or real number and n is non negative integers is called a polynomial function.

Linear polynomial: A polynomial with degree one is called linear polynomial. For e.g. 2x+5, 10x-6 etc

Quadratic polynomial: A polynomial with degree two is called quadratic polynomial. For e.g. $x^2 + 10x + 2$, $4x^2 + 3x - 5$, $x^2 + 4$ etc.

Cubic polynomial: A polynomial with degree three is called cubic polynomial. For e.g. $x^3 + 3x - 5$, $7x^2 - 6$ etc

11. ABSOLUTE VALUE FUNCTION OR MODULS FUNCTION: A function f is said to absolute value function if for any real value of x∈R is f(x) =|x| i.e. +R. Range of Absolute function is always [0,x)

12. EXPLICIT FUNCTION: A function f is said to be explicit, if the functions can be defined x and y separately. For e.g. $y = x^2$, $y = x^3 + 1$

13. IMPLICIT FUNCTION: A function f is said to be implicit if the variable x and y are not separable, also called homogenous function for e.g. $f(x) = g(x,y)$, $f(x) = \dfrac{x^2 + y^2}{x+y}$

14. DEMAND FUNCTION: By holding other determinates of demands as constant a consumer demand for a good as a function of its price so demand function is a inverse between price and quantity. We can represent equation, $q = f(p)$ or $f(p) = q$, where p- price of goods- quantity demand.

15. PRODUCTION FUNCTION: Production function is the relation between the firms production output and the material factors of production input. So, we simply can say input and output relation.

$Q(x) = f(L, K)$, Where $Q(x)$ - Output of Goods x,

L-Labour, a variable factor input   K-Capital, fixed factors

17. COST FUNCTION: A cost function is the study of functional relationship between cost and output; it shows the least cost combination of input corresponding to different level of output.

TC= TFC +TVC: TC-Total cost, TFC- Total fixed cost, TVC- Total variable cost.

18. REVENUE FUNCTION: The revenue of a firm is a sales receipts or money receipts from the sales of the product. TR= ARx Q, TR-Total Revenue, AR-Average revenue- Quantity.

19. RATIONAL FUNCTION: A function $f(x)$ which can be expressed in the form of $f(x) = \frac{g(x)}{h(x)}$ where $h(x) \neq 0$ is called rational function. The domain of rational function is all real values which provide $h(x) \neq 0$.

20. SUPPLY FUNCTION: Supply of a commodity is based on law of supply, i.e. direct relationship between price and supply of goods. So, supply function is the relationship between supplies of goods at different variation.

$Q^s x = c + d\ (p_x)$

$Q^s x$- Quantity supply of commodity x

P_x – Price of commodity-x

d- Change in quantity with respect of price

c- Constant supply when $(p_x = 0)$

21. UTILITY FUNCTION: It is the sum of total utility derived from the consumption of all the units of commodity in a particular income.

$TU = \sum MU$, where TU- Total utility, MU -TU_n- TU_{n-1}

22. CONSUMPTION FUNCTION: It is the relation between income of individual and his expenditure the total consumption of a firm is depending on the income as the income increases the consumption expenditure also increases.

If C is the consumption and I is income then consumption function is define by C=a +bI

Where C-Consumption, I-Income, a, b – constants

23. INVERSE FUNCTION: A function f is said to be inverse if f one-one and onto or bijective. Then, f: A→B

g: B→A, Which associate each element y belongs to a unique element x∈A, f(x)= y is called inverse of x.

f(x) = y,⇒ g(y)=x, The inverse of f is denote f^{-1}.

Composition function: Let f: A→B and g: B→C be two functions. Then the composition of f and g, denoted by gof, is defined as the function gof: A→C given by Gof(x) = g(f(x)), for all x A.

UNIT: VI

MATRICES AND DETERMINANTS

MATRICS:

A matrices' is a ordered rectangular array of number or functions is called entries of the matrices or elements of the matrices and the whole arrangement is called matrices.

[ ] or ( )

USES OF MATRICS:

It is useful mathematic technique of solve system of linear equation elements spread sheet, programs for personal computer, different uses in the business and science like budgeting, sales projection, cost estimation and various analyzing experiment. we can study the relation of more than two variables at a time .

ORDER OF MATRICS:

A matrix has an m-row and n-columns are called a matrix of order m x n order.

For example

$$=\begin{bmatrix} a_{11} & a_{12} \\ a_{31} & a_{32} \end{bmatrix}_{2X2}$$

$$= \begin{bmatrix} b_{11} & b_{12} & b_{13} \\ b_{21} & b_{22} & b_{23} \\ b_{31} & b_{31} & b_{33} \end{bmatrix} 3X3$$

$$= \begin{bmatrix} 4 & 5 & 7 \\ 1 & 3 & 6 \end{bmatrix} 2x3$$

Q.1. Write a various order of matrix having 36 elements and entries.

Sol. There are 9 different order of matrix found using the 36 elements I.e. 1x36, 2x18, 3x12, 4x9, 6x6, 9x4, 12x3, 13x2, 36x1.

Q. 2. Construct a matrix is 2x3 order which satisfy aij $= \frac{(i+j)^2}{2}$

Sol. let A= $\begin{bmatrix} a_{11} & a_{12} & a_{13} \\ a_{21} & a_{22} & a_{23} \end{bmatrix} 2x3$

$a_{ij} = \frac{(i+j)^2}{2}$ (Given)

$A11 = \frac{(1+1)^2}{2} = \frac{(2)^2}{2} = \frac{4}{2} = 2$

$A12 = \frac{(1+2)^2}{2} = \frac{9}{2}$

$A13 = \frac{(1+3)^2}{2} = \frac{16}{2} = 8$

$A21 = \frac{(2+1)^2}{2} = \frac{9}{2}$

$A22 = \frac{(2+2)^2}{2} = \frac{16}{2} = 8$

$A23 = \frac{(2+3)^2}{2} = \frac{25}{2}$

$$\therefore A = \begin{bmatrix} 2 & \frac{9}{2} & 8 \\ \frac{9}{2} & 8 & \frac{25}{2} \end{bmatrix}$$

Q 3. Construct a 3x2 matrix whose elements are given $a_{IJ} = \frac{1}{2} | i - 3j |$

Sol. let

$$A = \begin{bmatrix} a_{11} & a_{12} \\ a_{21} & a_{22} \\ a_{31} & a_{32} \end{bmatrix} 3x2 \begin{bmatrix} a_{11} & a_{12} \\ a_{21} & a_{22} \\ a_{31} & a_{32} \end{bmatrix}$$

$Aij = \frac{1}{2} |i-3j|$

$A_{11} = \frac{1}{2} |1-3(1)| = \frac{1}{2}|-2| = \frac{1}{2}x2$

$A_{11} = 1$

$A_{12} = \frac{1}{2} |1 - 3(2)| = \frac{1}{2}|1-6| = \frac{1}{2}x5$

$A_{12} = \frac{5}{2}$

$A_{21} = \frac{1}{2} |2-3(1)| = \frac{1}{2} |2-3| = \frac{1}{2}|-1| = \frac{1}{2}x1$

$A_{21} = \frac{1}{2}$

$A_{22} = \frac{1}{2} |2-3(1)| = \frac{1}{2} |2-6| = \frac{1}{2}|-4| = \frac{1}{2}x4$

$A_{22} = 2$

$A_{31} = \frac{1}{2} |3-3(1)| = \frac{1}{2} |3-3|$

$A_{32} = \frac{1}{2}x0$

$= 0$

$A_{32} = \frac{1}{2}|3-3(2)| \quad = \frac{1}{2}|3-6| \quad = \frac{1}{2}|-3|$

$A_{32} = \frac{3}{2}$

$$\therefore A = \begin{bmatrix} 1 & \frac{5}{2} \\ \frac{1}{2} & 2 \\ 0 & \frac{3}{2} \end{bmatrix} 3 \times 2$$

TYPES OF MATRICES

EQUAL MATRIX

If two matrix $A=[a_{ij}]$, $B=[b_{ij}]$ said to be equal if order of both A and B are equal. The elements of A = corresponding elements of B.

For example:

$\begin{bmatrix} 2 & 3 \\ 0 & 1 \end{bmatrix}$ and $\begin{bmatrix} 2 & 3 \\ 0 & 1 \end{bmatrix}$ are equal matrix.

Q.1. Find the value of a, b, c & d from the following equation.

$$\begin{bmatrix} 2a+b & a-2b \\ 5c-d & 4c+3d \end{bmatrix} = \begin{bmatrix} 4 & -3 \\ 11 & 24 \end{bmatrix}$$

Sol.

$$\begin{bmatrix} 2a+b & a-2b \\ 5c-d & 4c+3d \end{bmatrix} = \begin{bmatrix} 4 & -3 \\ 11 & 24 \end{bmatrix}$$

Comparing with corresponding elements

2a+b=4............(i)

a-2b=-3..............(ii)

5c-d=11.............(iii)

4c+3d=24.............(iv)

Solving (i) and (ii) and multiply by 2.

4a+2b=8

a-2b=-3

5a   =5

$a=\frac{5}{5}$

a=1

Put value of 'a' in equation (2)

1-2b = -3

-2b=-3-1

2b=4

$b=\frac{4}{2} = 2$

Solving (iii) & (iv)

5c-d=11.............(iii)

4c+3d=24...........(iv)

Multiply by 3 in equation (iii)

$3x \Rightarrow 15c\text{-}3d = 33$

4c+3d = 24

19c=57

$$c = \frac{57}{19}$$

c= 3

Put value of c in equation (iv)

4x3+3d=24

12+3d=24-12

3d=12

$$d = \frac{12}{3}$$

d=4

COLUMN MATRIX

A matrix is said to be a column matrix if it has only one column and any no. of rows.

For example:

$$\begin{bmatrix} 0 \\ 3 \\ 5 \end{bmatrix}$$

∴ General form of column matrix

A=$[aij]$ mx1

ROW MATRIX

A matrix is said to be row matrix, if it has only one row and any number columns.

For example:

[10 3 4]

∴ General form of Row matrix

$A = [a_{ij}]_{1xn}$

SQUARE MATRIX

A matrix in which no. of rows are equal to number of columns is said to be square matrix . i.e

$A = [a_{ij}]_{mxn} [m=n]$

For example: $\begin{bmatrix} 1 & 3 \\ 4 & 5 \end{bmatrix}_{2x2}$

DIAGONAL MATRIX

A square matrix is said to be a diagonal matrix if all the diagonal elements are non-zero numbers

For example;

$\begin{bmatrix} 5 & 0 \\ 0 & 7 \end{bmatrix} , \begin{bmatrix} 1 & 0 & 0 \\ 0 & 4 & 0 \\ 0 & 0 & 6 \end{bmatrix}$

SCALAR MATRIX

A Diagonal matrix is said to be a scalar matrix if all the diagonal elements are equal.

For example:

$\begin{bmatrix} 5 & 0 \\ 0 & 5 \end{bmatrix} \begin{bmatrix} 8 & 0 & 0 \\ 0 & 8 & 0 \\ 0 & 0 & 8 \end{bmatrix}$

UNITS MATRIX OR IDENTITY MATRIX

A square matrix image elements is the diagonal are all one and rest of all zeroes is called identity matrix we denote identity matrix by order and by in . it is simple written as I.

For example;

$I_1 = [1]$

$I_2 = \begin{bmatrix} 1 & 0 \\ 0 & 1 \end{bmatrix}$

$I_3 = \begin{bmatrix} 1 & 0 & 0 \\ 0 & 1 & 0 \\ 0 & 0 & 1 \end{bmatrix}$

ZERO MATRIX OR NULL MATRIX

A matrix is said to be zero matrix or null matrix if all its elements are zero.

For example

$[0] \quad \begin{bmatrix} 0 & 0 \\ 0 & 0 \end{bmatrix} \quad \begin{bmatrix} 0 & 0 & 0 \\ 0 & 0 & 0 \\ 0 & 0 & 0 \end{bmatrix}$

OPERATIONS ON MATRICES

ADDITION OF MATRICES

If two matrix A & B can add if the two matrix of the same order I.e. [A]mxn and [B]mxn then its sun 'c' matrix is given by the same order mxn I.e. $[c_{ij}]$mxn.

For example

Let $A = \begin{bmatrix} a_{11} & a_{12} & a_{13} \\ a_{21} & a_{22} & a_{23} \end{bmatrix}_{2 \times 3}$

$$B = \begin{bmatrix} b_{11} & b_{12} & b_{13} \\ b_{21} & b_{22} & b_{23} \end{bmatrix}_{2 \times 3}$$

Sol. $A+B = \begin{pmatrix} a_{11} & a_{12} & a_{13} \\ a_{21} & a_{22} & a_{23} \end{pmatrix}_{2 \times 3} + \begin{pmatrix} b_{11} & b_{12} & b_{13} \\ b_{21} & b_{22} & b_{23} \end{pmatrix}_{2 \times 3}$

$$\therefore C = \begin{bmatrix} a_{11}+b_{11} & a_{12}+b_{12} & a_{13}+b_{13} \\ a_{21}+b_{21} & a_{22}+b_{22} & b_{23}+b_{23} \end{bmatrix}_{2 \times 3}$$

DIFFERENCE OF TWO MATRICES

Let $A=[a_{ij}]$ $B=[b_{ij}]$ are the two matrices of the same order (mxn) then the different of A&B is D. therefore $D=[d_{ij}]_{mxn}$ Here $d_{ij}=a_{ij}-b_{ij}$

Q 1. If $A=\begin{bmatrix} 1 & 2 & 3 \\ 2 & 3 & 1 \end{bmatrix}$, $B=\begin{bmatrix} 1 & -1 & 3 \\ -1 & 0 & 2 \end{bmatrix}$

$2A-B=2\begin{bmatrix} 1 & 2 & 3 \\ 2 & 3 & 1 \end{bmatrix} - \begin{bmatrix} 1 & -1 & 3 \\ -1 & 0 & 2 \end{bmatrix}$

$=\begin{bmatrix} 2 & 4 & 6 \\ 4 & 6 & 2 \end{bmatrix} - \begin{bmatrix} 3 & -1 & 3 \\ -1 & 0 & 2 \end{bmatrix}$

$=\begin{bmatrix} 2-3 & 4+1 & 6-3 \\ 4+1 & 6-0 & 2-2 \end{bmatrix}$

$\therefore 2A-B=\begin{bmatrix} -1 & 5 & 3 \\ 5 & 6 & 0 \end{bmatrix}$

Q. 2 If $A=\begin{bmatrix} 2 & 4 \\ 5 & 6 \end{bmatrix}$, $B=\begin{bmatrix} -1 & 7 \\ -3 & 8 \end{bmatrix}$ find 5A-3A.

Sol. $A=\begin{bmatrix} 2 & 4 \\ 5 & 6 \end{bmatrix}$, $B=\begin{bmatrix} -1 & 7 \\ -3 & 8 \end{bmatrix}$

$\therefore 5A-3A=5\begin{bmatrix} 2 & 4 \\ 5 & 6 \end{bmatrix} - 3\begin{bmatrix} -1 & 7 \\ -3 & 3 \end{bmatrix}$

$$= \begin{bmatrix} 10 & 20 \\ 25 & 30 \end{bmatrix} - \begin{bmatrix} -1 & 21 \\ -9 & 24 \end{bmatrix}$$

$$= \begin{bmatrix} 10+3 & 20-21 \\ 25+9 & 30-24 \end{bmatrix}$$

$$\therefore 5A-3B = \begin{bmatrix} 13 & -1 \\ 34 & 6 \end{bmatrix}$$

Q. 3 $A = \begin{bmatrix} 2 & 0 & -1 \\ 4 & 1 & 4 \end{bmatrix}$, $B = \begin{bmatrix} 1 & 1 & 6 \\ 1 & 3 & 4 \end{bmatrix}$ $C = \begin{bmatrix} -7 & 8 & 1 \\ 0 & 1 & 4 \end{bmatrix}$ find A+2B-3C

Sol. $A = \begin{bmatrix} 2 & 0 & -1 \\ 4 & 1 & 4 \end{bmatrix}$, $B = \begin{bmatrix} 1 & 1 & 6 \\ 1 & 3 & 4 \end{bmatrix}$ $C = \begin{bmatrix} -7 & 8 & 1 \\ 0 & 1 & 4 \end{bmatrix}$

$$\therefore A+2B-3C = \begin{bmatrix} 2 & 0 & -1 \\ 4 & 1 & 4 \end{bmatrix} + 2\begin{bmatrix} 1 & 1 & 6 \\ 1 & 3 & 4 \end{bmatrix} - 3\begin{bmatrix} -7 & 8 & 1 \\ 0 & 1 & 4 \end{bmatrix}$$

$$= \begin{bmatrix} 2 & 0 & -1 \\ 4 & 1 & 4 \end{bmatrix} + \begin{bmatrix} 2 & -2 & 12 \\ 2 & 6 & 8 \end{bmatrix} - \begin{bmatrix} -21 & 24 & 3 \\ 0 & 3 & 12 \end{bmatrix}$$

$$= \begin{bmatrix} 2+2 & 0-2 & -1+12 \\ 4+2 & 1+6 & 4+8 \end{bmatrix} - \begin{bmatrix} 21 & 24 & 3 \\ 0 & 3 & 12 \end{bmatrix}$$

$$= \begin{bmatrix} 4 & -2 & 11 \\ 6 & 7 & 12 \end{bmatrix} - \begin{bmatrix} 21 & 24 & 3 \\ 0 & 3 & 12 \end{bmatrix}$$

$$= \begin{bmatrix} 4+21 & -2-24 & -11-3 \\ 6-0 & 7-3 & 12-12 \end{bmatrix}$$

$$\therefore A+2B-3C = \begin{bmatrix} 25 & -26 & 8 \\ 6 & 4 & 0 \end{bmatrix}$$

Q. 4 Find x and y; $x+y = \begin{bmatrix} 7 & 0 \\ 2 & 5 \end{bmatrix}$ $x-y = \begin{bmatrix} 3 & 0 \\ 0 & 3 \end{bmatrix}$

Sol. $X+y = \begin{bmatrix} 7 & 0 \\ 2 & 5 \end{bmatrix}$.........(i)

$x-y = \begin{bmatrix} 3 & 0 \\ 0 & 3 \end{bmatrix}$............(ii)

Adding $2x=\begin{bmatrix} 7 & 0 \\ 2 & 5 \end{bmatrix}+\begin{bmatrix} 3 & 0 \\ 0 & 3 \end{bmatrix}$

$$2x=\begin{bmatrix} 10 & 0 \\ 2 & 8 \end{bmatrix}$$

$$X=\frac{1}{2}\begin{bmatrix} 10 & 0 \\ 2 & 8 \end{bmatrix}$$

$$X=\begin{bmatrix} 5 & 0 \\ 1 & 4 \end{bmatrix}$$

Substralt (ii) from (i)

$$2y=\begin{bmatrix} 7 & 0 \\ 2 & 5 \end{bmatrix}-\begin{bmatrix} 3 & 0 \\ 0 & 3 \end{bmatrix}$$

$$2y=\begin{bmatrix} 4 & 0 \\ 2 & 2 \end{bmatrix}$$

$$Y=\frac{1}{2}\begin{bmatrix} 4 & 0 \\ 2 & 2 \end{bmatrix}$$

$$Y=\begin{bmatrix} 2 & 0 \\ 1 & 1 \end{bmatrix}$$

Q. 5 Find 'x' & 'y'

$$2x+3y=\begin{bmatrix} 2 & 3 \\ 4 & 0 \end{bmatrix}$$

$$3x+2y=\begin{bmatrix} 2 & -2 \\ -1 & 5 \end{bmatrix}$$

Sol. $2x+3y=\begin{bmatrix} 2 & 3 \\ 4 & 0 \end{bmatrix}$.............(i)

$3x+2y=\begin{bmatrix} 2 & -2 \\ -1 & 5 \end{bmatrix}$...............(ii)

Equation –(i) multiply by $3\Rightarrow$

$$\Rightarrow 6x+9y=\begin{bmatrix} 6 & 9 \\ 12 & 0 \end{bmatrix}\ldots\ldots\ldots(iii)$$

Equation –(ii) multiply by 2 $\Rightarrow$

$$\Rightarrow 6x+4y=\begin{bmatrix} 4 & -4 \\ -2 & 10 \end{bmatrix}\ldots\ldots\ldots(iv)$$

Equation -(iii)-(iv)$\Rightarrow$

$$\Rightarrow 6x+9y=\begin{bmatrix} 6 & 9 \\ 12 & 0 \end{bmatrix}$$

$$\Rightarrow 6x+4y=\begin{bmatrix} 4 & -4 \\ -2 & 10 \end{bmatrix}$$

$$5y=\begin{bmatrix} 6 & 9 \\ 12 & 0 \end{bmatrix}-\begin{bmatrix} 4 & -4 \\ -2 & 10 \end{bmatrix}$$

$$5y=\begin{bmatrix} 2 & 13 \\ 14 & 10 \end{bmatrix}$$

$$y=\frac{1}{5}\begin{bmatrix} 2 & 13 \\ 14 & -10 \end{bmatrix}$$

$$y=\begin{bmatrix} \frac{2}{5} & \frac{13}{5} \\ \frac{14}{5} & -2 \end{bmatrix}$$

put value of y$=\begin{bmatrix} \frac{2}{5} & \frac{13}{5} \\ \frac{14}{5} & -2 \end{bmatrix}$ in equation –(i)

$$\Rightarrow 2x+\begin{bmatrix} \frac{2}{5} & \frac{13}{5} \\ \frac{14}{5} & -2 \end{bmatrix}=\begin{bmatrix} 2 & 3 \\ 4 & 0 \end{bmatrix}$$

$$\Rightarrow 2x + \begin{bmatrix} \frac{6}{5} & \frac{39}{5} \\ \frac{42}{5} & -6 \end{bmatrix} = \begin{bmatrix} 2 & 3 \\ 4 & 0 \end{bmatrix}$$

$$\Rightarrow 2x = \begin{bmatrix} 2 & 3 \\ 4 & 0 \end{bmatrix} - \begin{bmatrix} \frac{6}{5} & \frac{39}{5} \\ \frac{42}{5} & -6 \end{bmatrix}$$

$$\Rightarrow 2x \Rightarrow \begin{bmatrix} 2 - \frac{6}{5} & 3 - \frac{39}{5} \\ 4 - \frac{42}{5} & 0 - 6 \end{bmatrix}$$

$$\Rightarrow 2x = \begin{bmatrix} \frac{4}{5} & \frac{24}{5} \\ \frac{22}{5} & -6 \end{bmatrix}$$

$$\Rightarrow x = \frac{1}{2} \begin{bmatrix} \frac{4}{5} & \frac{24}{5} \\ \frac{-22}{5} & -6 \end{bmatrix}$$

$$\Rightarrow x = \begin{bmatrix} \frac{2}{5} & \frac{-12}{5} \\ \frac{-11}{5} & -3 \end{bmatrix}$$

Q.6. Find value of X and y from the following equations.

$$2\begin{bmatrix} x & 5 \\ 7 & y-3 \end{bmatrix} + \begin{bmatrix} 3 & -4 \\ 1 & 2 \end{bmatrix} = \begin{bmatrix} 7 & 6 \\ 15 & 14 \end{bmatrix}$$

Sol: $2\begin{bmatrix} x & 5 \\ 7 & y-3 \end{bmatrix} + \begin{bmatrix} 3 & -4 \\ 1 & 2 \end{bmatrix} = \begin{bmatrix} 7 & 6 \\ 15 & 14 \end{bmatrix}$

$$\Rightarrow \begin{bmatrix} 2x & 10 \\ 14 & 2y-6 \end{bmatrix} = \begin{bmatrix} 7 & 6 \\ 15 & 14 \end{bmatrix} = \begin{bmatrix} 3 & -4 \\ 1 & 2 \end{bmatrix}$$

$$\Rightarrow \begin{bmatrix} 2x & 10 \\ 14 & 2y-6 \end{bmatrix} = \begin{bmatrix} 7-3 & 6+4 \\ 15-1 & 14-2 \end{bmatrix}$$

$$\Rightarrow \begin{bmatrix} 2x & 10 \\ 14 & 2y-6 \end{bmatrix} = \begin{bmatrix} 4 & 10 \\ 14 & 12 \end{bmatrix}$$

Comparing with like corresponding

$\Rightarrow 2x=4$

$X=\frac{4}{2} \Rightarrow 2$

$\Rightarrow 2y-6=12$

$2y=12+6$

$2y=18$

$Y=\frac{18}{2}=9$

$\therefore$ x=2 and y=9

Q. 7 If A=$\begin{bmatrix} 8 & 0 \\ 4 & -2 \\ 3 & 6 \end{bmatrix}$, B=$\begin{bmatrix} 2 & -2 \\ 4 & 2 \\ -5 & 1 \end{bmatrix}$ find x if 2A+3X=5B

Sol. A=$\begin{bmatrix} 8 & 0 \\ 4 & -2 \\ 3 & 6 \end{bmatrix}$, B=$\begin{bmatrix} 2 & -2 \\ 4 & 2 \\ -5 & 1 \end{bmatrix}$

$\Rightarrow 2A+3X=5B$

$$\Rightarrow \begin{bmatrix} 8 & 0 \\ 4 & -2 \\ 3 & 6 \end{bmatrix} + 3x = 5\begin{bmatrix} 2 & -2 \\ 4 & 2 \\ -5 & 1 \end{bmatrix}$$

$$\Rightarrow \begin{bmatrix} 16 & 0 \\ 8 & -4 \\ 6 & 12 \end{bmatrix} + 3x = \begin{bmatrix} 10 & -10 \\ 20 & 10 \\ -25 & 5 \end{bmatrix}$$

$$\Rightarrow 3x = \begin{bmatrix} 10 & -10 \\ 20 & 10 \\ -25 & 5 \end{bmatrix} - \begin{bmatrix} 16 & 0 \\ 8 & -4 \\ 6 & 12 \end{bmatrix}$$

$$\Rightarrow 3x = \begin{bmatrix} 10-16 & -10-0 \\ 20-8 & 10+4 \\ -25-6 & 5-12 \end{bmatrix}$$

$$\Rightarrow 3x = \begin{bmatrix} -6 & -10 \\ 12 & 14 \\ -31 & -7 \end{bmatrix}$$

$$\Rightarrow x = \frac{1}{3} \begin{bmatrix} 6 & 10 \\ 12 & 14 \\ -31 & -7 \end{bmatrix}$$

$$\Rightarrow x = \begin{bmatrix} -2 & \frac{-10}{3} \\ 4 & 7 \\ \frac{-31}{3} & \frac{-7}{3} \end{bmatrix}$$

MATRICES MULTIPLICATION

The product two matrix 'A' and 'B' are defined if the number of columns of 'A' is equal to the number of rows of 'B'.

Let $A = [a_{IJ}]_{mxp}$ $B = [b_{ij}]_{pxn}$ then its result $c = [c_{ij}]_{mxn}$

Q.1 Find AB, if $A = \begin{bmatrix} 0 & -1 \\ 0 & 2 \end{bmatrix}$ $B = \begin{bmatrix} 3 & 5 \\ 0 & 0 \end{bmatrix}$

Sol. $AB = \begin{bmatrix} 0 & -1 \\ 0 & 2 \end{bmatrix} 2x2 \begin{bmatrix} 3 & 5 \\ 0 & 0 \end{bmatrix} 2x2$

$$= A = \begin{bmatrix} (0X3) + (-1X0) & (0X5 + (-1X0) \\ (0X3) + (2X0) & (0X5) + (2X0) \end{bmatrix}$$

$$= A = \begin{bmatrix} 0 - 0 & 0 + 0 \\ 0 + 0 & 0 + 0 \end{bmatrix}$$

$$AB = \begin{bmatrix} 0 & 0 \\ 0 & 0 \end{bmatrix} 2x2$$

Q .2 Find A = $\begin{bmatrix} -3 & 4 \\ 1 & 3 \end{bmatrix}$ B = $\begin{bmatrix} 1 & -2 \\ 6 & 4 \end{bmatrix}$

Sol. A = $\begin{bmatrix} -3 & 4 \\ 1 & 3 \end{bmatrix}$ 2x2 B = $\begin{bmatrix} 1 & -2 \\ 6 & 4 \end{bmatrix}$ 2x2

$$\Rightarrow \begin{bmatrix} -3 & 4 \\ 1 & 3 \end{bmatrix} \begin{bmatrix} 1 & -2 \\ 6 & 4 \end{bmatrix}$$

$$\Rightarrow \begin{bmatrix} (-3X1) + (4X6) & (-3X-2)(4X4) \\ (1X1) + (3X6) & (1X-2)(3X4) \end{bmatrix}$$

$$\Rightarrow \begin{bmatrix} -3 + 24 & -6 + 16 \\ 1 + 18 & -2 + 12 \end{bmatrix}$$

$$\Rightarrow \begin{bmatrix} 21 & 22 \\ 19 & 10 \end{bmatrix}$$

Q 3. Find A = $\begin{bmatrix} 1 & 0 & 4 \\ 1 & -1 & 2 \\ 1 & 3 & 1 \end{bmatrix}$ B = $\begin{bmatrix} 6 & 1 & 2 \\ 2 & 0 & 1 \\ 4 & 1 & 4 \end{bmatrix}$

Sol. A = $\begin{bmatrix} 1 & 0 & 4 \\ 1 & -1 & 2 \\ 1 & 3 & 1 \end{bmatrix}$ 3x3 B = $\begin{bmatrix} 6 & 1 & 2 \\ 2 & 0 & 1 \\ 4 & 1 & 4 \end{bmatrix}$ 3x3

=

$$\begin{bmatrix} (1X6) + (0X2) + (4X4) & (1X1) + (0X0) + (4X1) & (1X2) + (0X1) + (4X4) \\ (1X6) + (-1X2) + (2X4) & (1X1) + (-1X0) + (2X1) & (1X2) + (-1X1) + (2X4) \\ (1X6) + (3X2) + (1X4) & (1X1) + (3X0) + (1X1) & (1X2) + (3X1) + (1X4) \end{bmatrix}$$

$$= \begin{bmatrix} 6+0+16 & 1+0+4 & 2+0+16 \\ 6-2+8 & 1-0+2 & 2-1+8 \\ 6+6+4 & 1+0+1 & 2+3+4 \end{bmatrix}$$

$$= \begin{bmatrix} 2 & 5 & 18 \\ 16 & 3 & 9 \\ 16 & 2 & 9 \end{bmatrix}$$

Q 4. A= $\begin{bmatrix} 1 & 2 & 3 \\ 3 & -2 & 1 \\ 4 & 2 & 1 \end{bmatrix}$ show that A^3-23A-40I=0

Sol. A^2=A,A

$$= \begin{bmatrix} 1 & 2 & 3 \\ 3 & -2 & 1 \\ 4 & 2 & 1 \end{bmatrix} \begin{bmatrix} 1 & 2 & 3 \\ 3 & -2 & 1 \\ 4 & 2 & 1 \end{bmatrix}$$

$$= \begin{bmatrix} 19+12+32 & 38-8+16 & 57+4+8 \\ 1+36++32 & 2-24+16 & 3+12+8 \\ 14+18+60 & 28+12+30 & 42+6+15 \end{bmatrix}$$

$$= \begin{bmatrix} 63 & 46 & 69 \\ 69 & -6 & 23 \\ 92 & 46 & 63 \end{bmatrix}$$

L.H.S. A^3 -23 A -40I = 0

$$\Rightarrow \begin{bmatrix} 63 & 46 & 69 \\ 69 & -6 & 23 \\ 92 & 46 & 63 \end{bmatrix} -23 \begin{bmatrix} 1 & 2 & 3 \\ 3 & 2 & 1 \\ 4 & 2 & 1 \end{bmatrix} -40 \begin{bmatrix} 1 & 0 & 0 \\ 0 & 1 & 0 \\ 0 & 0 & 1 \end{bmatrix}$$

$$\Rightarrow \begin{bmatrix} 63 & 46 & 69 \\ 69 & -6 & 23 \\ 92 & 46 & 63 \end{bmatrix} - \begin{bmatrix} 23 & 46 & 69 \\ 69 & -46 & 23 \\ 92 & 46 & 23 \end{bmatrix} -40 \begin{bmatrix} 1 & 0 & 0 \\ 0 & 1 & 0 \\ 0 & 0 & 1 \end{bmatrix}$$

$$\Rightarrow \begin{bmatrix} 63-23 & 46-46 & 69-69 \\ 69-69 & -6+46 & 23-23 \\ 92-92 & 46-46 & 63-23 \end{bmatrix} - \begin{bmatrix} 40 & 0 & 0 \\ 0 & 40 & 0 \\ 0 & 0 & 40 \end{bmatrix} =0$$

$$\Rightarrow \begin{bmatrix} 40 & 0 & 0 \\ 0 & 40 & 0 \\ 0 & 0 & 40 \end{bmatrix} - \begin{bmatrix} 40 & 0 & 0 \\ 0 & 40 & 0 \\ 0 & 0 & 40 \end{bmatrix} = 0$$

$$\Rightarrow \begin{bmatrix} 0 & 0 & 0 \\ 0 & 0 & 0 \\ 0 & 0 & 0 \end{bmatrix}$$

L.H.S=R.H.S

Q5. In a legislative assembly election a political group him a public relation to promote its candidate in three wags telephone calls and letters. The cost per contact (in price paise) is given a matrix $A = \begin{bmatrix} 40 \\ 100 \\ 50 \end{bmatrix}$

There number of contact of each type made in two cities X and y is given by matrix $B = \begin{bmatrix} 1000 & 5000 & 5000 \\ 3000 & 1000 & 10000 \end{bmatrix}$

Find the total amount spent by the group in two cities X and Y?

Sol. To find the total expenditure we need to find the product A&B

$$\therefore A = \begin{bmatrix} 40 \\ 100 \\ 50 \end{bmatrix} \quad B = \begin{bmatrix} 1000 & 5000 & 5000 \\ 3000 & 1000 & 10000 \end{bmatrix}$$

$$AB = \begin{bmatrix} 40 \\ 100 \\ 50 \end{bmatrix} \begin{bmatrix} 1000 & 5000 & 5000 \\ 3000 & 1000 & 10000 \end{bmatrix}$$

Change AB into BA

$$BA=\begin{bmatrix}1000 & 5000 & 5000\\3000 & 1000 & 10000\end{bmatrix}2x1\begin{bmatrix}40\\100\\50\end{bmatrix}3x1$$

$$=\begin{bmatrix}40000+50000+250000\\120000+100000+500000\end{bmatrix}$$

$$=\begin{bmatrix}340000\\720000\end{bmatrix}$$

$$=\begin{bmatrix}3400\\7200\end{bmatrix}$$

∴ Total expenditure $=34000+72000$

$$=106000$$

Q 6. If $A=\begin{bmatrix}1 & 2 & 2\\2 & 1 & 2\\2 & 2 & 1\end{bmatrix}$ show that A^2-4A-51=0. Hence find the inverse of A.

Sol. A^2-4A-4-51=0

Multiply by A^2,A^1 -4A.A^{-1}-51.A^{-1}=0

$A(AXA^{-1})$-4I-5A^{-1}=0

A-4I-5A^{-1}=0

-5A^{-1}=4I-5A

$$-5A^{-1}=4\begin{bmatrix}1 & 0 & 0\\0 & 1 & 0\\0 & 0 & 1\end{bmatrix}-\begin{bmatrix}1 & 2 & 2\\2 & 1 & 2\\2 & 2 & 1\end{bmatrix}$$

$$-5A^{-1}=\begin{bmatrix}4 & 0 & 0\\0 & 4 & 0\\0 & 0 & 4\end{bmatrix}-\begin{bmatrix}1 & 2 & 2\\2 & 1 & 2\\2 & 2 & 1\end{bmatrix}$$

$$-5A^{-1} = \begin{bmatrix} -3 & -2 & -2 \\ -2 & 3 & -2 \\ -2 & -2 & 3 \end{bmatrix}$$

$$A^{-1} = \frac{1}{5}\begin{bmatrix} -3 & -2 & -2 \\ -2 & 3 & -2 \\ -2 & -2 & 3 \end{bmatrix}$$

$$A^{-1} = \begin{bmatrix} \frac{-3}{5} & \frac{-2}{5} & \frac{2}{5} \\ \frac{2}{5} & \frac{-3}{5} & \frac{2}{5} \\ \frac{2}{5} & \frac{2}{5} & \frac{-3}{5} \end{bmatrix}$$

Q.7. Show that $A = \begin{bmatrix} 2 & -1 & 1 \\ -1 & 2 & -1 \\ 1 & -1 & 2 \end{bmatrix}$ satisfies the equation $A^3 - 6A^2 + 9A - 4I = 0$ and hence find A^{-1}

Sol. $A^2 = A.A$

$$= \begin{bmatrix} 2 & -1 & 1 \\ -1 & 2 & -1 \\ 1 & -1 & 2 \end{bmatrix}.\begin{bmatrix} 2 & -1 & 1 \\ -1 & 2 & -1 \\ 1 & -1 & 2 \end{bmatrix}$$

$$= \begin{bmatrix} 4+1+1 & -2-2-1 & 2+1+2 \\ -2-2-1 & 1+4+1 & -2-2-2 \\ 2+1+2 & -2-2-2 & 1+1+4 \end{bmatrix}$$

$$A^2 = \begin{bmatrix} 6 & -5 & 5 \\ -5 & 6 & -5 \\ 5 & -5 & 6 \end{bmatrix}$$

$A^3 = A^2.A$

$$\begin{bmatrix} 6 & -5 & 5 \\ -5 & 6 & -5 \\ 5 & -5 & 6 \end{bmatrix} \begin{bmatrix} 2 & -1 & 1 \\ -1 & 2 & -1 \\ 1 & -1 & 2 \end{bmatrix}$$

$$= \begin{bmatrix} 12+5+5 & -6-10-5 & 6+5+10 \\ -10-6-5 & 5+12+5 & -5-6-10 \\ 10+5+6 & -5-10-6 & 5+5+12 \end{bmatrix}$$

$$A^3 = \begin{bmatrix} 22 & -21 & 21 \\ -21 & 22 & -21 \\ 21 & -21 & 22 \end{bmatrix}$$

$A^3 - 6A^2 + 9A - 4I = 0$

Multiply by A^{-1} for both side

$\Rightarrow A^3.A^{-1} - 6A^2.A^{-1} + 9A.A^{-1} - 4I.A^{-1} = 0$

$\Rightarrow A^2(A.A^{-1}) - 6A(A.A^{-1}) + 9I - 4A^{-1} = 0$

$\Rightarrow A^2 - 6A + 9I - 4A^{-1} = 0$

$$\Rightarrow \begin{bmatrix} 6 & -5 & 5 \\ -5 & 6 & -5 \\ 5 & -5 & 6 \end{bmatrix} -$$
$$6\begin{bmatrix} 2 & -1 & 1 \\ -1 & 2 & -1 \\ 1 & -1 & 2 \end{bmatrix} + 9\begin{bmatrix} 1 & 0 & 0 \\ 0 & 1 & 0 \\ 0 & 0 & 1 \end{bmatrix} = 4A^{-1}$$

$$\Rightarrow \begin{bmatrix} 6 & -5 & 5 \\ -5 & 6 & -5 \\ 5 & -5 & 6 \end{bmatrix} -$$
$$6\begin{bmatrix} 12 & -6 & 6 \\ -6 & 12 & -6 \\ 6 & -6 & 12 \end{bmatrix} + 9\begin{bmatrix} 9 & 0 & 0 \\ 0 & 9 & 0 \\ 0 & 0 & 9 \end{bmatrix} = 4A^{-1}$$

$$\Rightarrow \begin{bmatrix} 6-12+9 & -5+6+0 & 5-6+0 \\ -5+6+0 & 6-12+9 & -5+6+0 \\ 5-6+0 & -5+6+0 & 6-12+9 \end{bmatrix} = 4A^{-1}$$

$$\Rightarrow \begin{bmatrix} -3 & 1 & -1 \\ 1 & -3 & 1 \\ -1 & -1 & -3 \end{bmatrix} = 4A^{-1}$$

$$\Rightarrow 4A^{-1}=\frac{1}{4}\begin{bmatrix} -3 & 1 & -1 \\ 1 & -3 & 1 \\ -1 & 1 & -3 \end{bmatrix}$$

$$\Rightarrow \begin{bmatrix} \frac{-3}{4} & \frac{1}{4} & \frac{-1}{4} \\ \frac{1}{4} & \frac{-3}{4} & \frac{1}{4} \\ \frac{-1}{4} & \frac{1}{4} & \frac{-3}{4} \end{bmatrix}$$

PROBLEMS FOR PRACTICE

Q.1 Find x,y ,z and t if $2\begin{bmatrix} x & z \\ y & t \end{bmatrix}+3\begin{bmatrix} 1 & -1 \\ 0 & 2 \end{bmatrix}=3\begin{bmatrix} 3 & 5 \\ 4 & 6 \end{bmatrix}$
(Ans:x=3,y=6,z=9,t=6)

Q.2 If A= $\begin{bmatrix} 2 & 1 & 1 \\ 3 & -1 & 0 \\ 0 & 2 & 4 \end{bmatrix}$ B= $\begin{bmatrix} 9 & 7 & -1 \\ 3 & 5 & 4 \\ 2 & 1 & 6 \end{bmatrix}$ and C= $\begin{bmatrix} 2 & -4 & 3 \\ 1 & -1 & 0 \\ 9 & 4 & 5 \end{bmatrix}$ verify that (A+B)+C=A+(B+C)

Q.3 Find a Matrix X such that 2A+B+X=0, Where

$A=\begin{bmatrix} -1 & 2 \\ 3 & 4 \end{bmatrix}$ B= $\begin{bmatrix} 3 & -2 \\ 1 & 5 \end{bmatrix}$ (Ans: x= $\begin{bmatrix} -1 & -2 \\ -7 & -13 \end{bmatrix}$

Q.4 Find the product of A and B, where A= $\begin{bmatrix} 5 & -1 \\ 6 & 7 \end{bmatrix}$ and B = $\begin{bmatrix} 2 & 1 \\ 3 & 4 \end{bmatrix}$

Q.5 If $\begin{bmatrix} 1 & -1 & x \end{bmatrix}\begin{bmatrix} 0 & 1 & -1 \\ 2 & 1 & 3 \\ 1 & 1 & 1 \end{bmatrix}\begin{bmatrix} 0 \\ 1 \\ 1 \end{bmatrix}=0$, Find X. (Ans: x=2)

Q.6 Three shop keepers A, B and C go to a store to buy stationary. A purchase 12 dozen notebooks, 5 dozen pens and 6 dozen pencils. B purchases 10 dozen notebooks 6 dozen pens and 7 dozen pencils' purchase 11 dozen notebooks,13 dozen pens and 8 dozen pencils. A notebook cost 40paise, a pen costs Rs1.25 and a pencil cost 35 paisa. Use matrix multiples to calculate individual's bill.

Q.7 If $A = \begin{bmatrix} 1 & 0 & -2 \\ 3 & -1 & 0 \\ -2 & 1 & 1 \end{bmatrix}$ $B = \begin{bmatrix} 0 & 5 & -4 \\ -2 & 1 & 3 \\ -1 & 0 & 2 \end{bmatrix}$
$C = \begin{bmatrix} 1 & 5 & 2 \\ -1 & 1 & 0 \\ 0 & -1 & 1 \end{bmatrix}$, verify that A (B-C) =AB-AC.

Q.8 If $A = \begin{bmatrix} 3 & 2 & 0 \\ 1 & 4 & 0 \\ 0 & 0 & 5 \end{bmatrix}$, Show that $A^2 - 7A + 10I_3 = 0$.

Transpose of a Matrix

Definition: Let A $[a_{ij}]$ be an mxn matrix. Then the transpose of 'A' denoted by A^Tor A' is an nxm matrix.

Q. If A $= \begin{bmatrix} -1 \\ 2 \\ 3 \end{bmatrix}$, B=[−2 −1 −4], verify that $(AB)^T = B^T A^T$

Sol. We have,

$A = \begin{bmatrix} -1 \\ 2 \\ 3 \end{bmatrix}$, B=[−2 −1 −4]

∴ AB= $\begin{bmatrix} -1 \\ 2 \\ 3 \end{bmatrix}$ [−2 −1 −4].

$$= \begin{bmatrix} 2 & 1 & 4 \\ -4 & -2 & -8 \\ -6 & -3 & -12 \end{bmatrix}$$

...... (i)

$$(AB)^T = \begin{bmatrix} 2 & -4 & -6 \\ 1 & -2 & -3 \\ 4 & -8 & -12 \end{bmatrix}$$

Also,

$$B^T A^T = \begin{bmatrix} -2 \\ -1 \\ -4 \end{bmatrix} = \begin{bmatrix} 1 & 2 & 3 \end{bmatrix}$$

$$= \begin{bmatrix} 2 & -4 & -6 \\ 1 & -2 & -3 \\ 4 & -8 & -12 \end{bmatrix} \quad ....(ii)$$

From (i) and (ii), we observed that

$(AB)^T = B^T A^T$

SYMMETRIC AND SKEW–SYMMETRIC MATRICES

Symmetric Matrix:

A square matrix A $= [a_{ij}]$ is called a symmetric matrix, if $a_{ij} = a_{ji}$ for all I,j.

For example, the matrix A $= \begin{bmatrix} 3 & -1 & 1 \\ -1 & 2 & 5 \\ 1 & 5 & -2 \end{bmatrix}$ is a symmetric, because $a_{12} = -1 = a_{21}$

$A_{13} = a_{31} = 1$

Thus, a square matrix A is a symmetric matrix if $A^T = A$

Skew−Symmetric Matrix

A square matrix A $=\begin{bmatrix} a_{ij} \end{bmatrix}$ is called a skew-symmetric matrix, if $a_{ij}=-a_{ji}$ for all I,j.

For example, A $=\begin{bmatrix} 0 & 2 & -3 \\ -2 & 0 & 5 \\ 3 & -5 & 0 \end{bmatrix}$

Here, $a_{12}=-a_{21}, a_{13}=a_{31}$ etc.

Thus, a square matrix A is skew – symmetric matrix if $A^T=-A$

We can express a matrix as a sum of symmetric and a skew-symmetric as follows

Symmetric of a matrix A= $A+A^T$

Skew symmetric of a matrix A= $A-A^T$

So, we can express it A=$\frac{1}{2}(A+A^T)+\frac{1}{2}(A-A^T)$

Q. Express the matrix A=$\begin{bmatrix} 3 & 2 & 3 \\ 4 & 5 & 3 \\ 2 & 4 & 5 \end{bmatrix}$ as a sum of a symmetric and a skew-symmetric matrix.

Sol. We have,

$A=\begin{bmatrix} 3 & 2 & 3 \\ 4 & 5 & 3 \\ 2 & 4 & 5 \end{bmatrix} \Rightarrow A^T=\begin{bmatrix} 3 & 4 & 2 \\ 2 & 5 & 4 \\ 3 & 3 & 5 \end{bmatrix}$

So, $A+A^T=\begin{bmatrix} 3 & 2 & 3 \\ 4 & 5 & 3 \\ 2 & 4 & 5 \end{bmatrix} + \begin{bmatrix} 3 & 4 & 2 \\ 2 & 5 & 4 \\ 3 & 3 & 5 \end{bmatrix}$

$$= \begin{bmatrix} 6 & 6 & 5 \\ 6 & 10 & 7 \\ 5 & 7 & 10 \end{bmatrix}$$

And, $A - A^T = \begin{bmatrix} 3 & 2 & 3 \\ 4 & 5 & 3 \\ 2 & 4 & 5 \end{bmatrix} - \begin{bmatrix} 3 & 4 & 2 \\ 2 & 5 & 4 \\ 3 & 3 & 5 \end{bmatrix}$

$$= \begin{bmatrix} 0 & -2 & 1 \\ 2 & 0 & -1 \\ -1 & 1 & 0 \end{bmatrix}$$

Let $P = \frac{1}{2}(A + A^T) = \begin{bmatrix} 3 & 3 & \frac{5}{2} \\ 3 & 5 & \frac{7}{2} \\ \frac{5}{2} & \frac{7}{2} & 5 \end{bmatrix}$

$$Q = \frac{1}{2}(A - A^T) = \begin{bmatrix} 0 & -1 & \frac{1}{2} \\ 1 & 0 & \frac{-1}{2} \\ \frac{-1}{2} & \frac{1}{2} & 0 \end{bmatrix}$$

∴ A = P+Q, Here P is asymmetric and Q is skew-symmetric.

PROBLEMS FOR PRACTICE

Q.1 If $A = \begin{bmatrix} 1 & -1 & 0 \\ 2 & 1 & 3 \\ 1 & 2 & 1 \end{bmatrix}$ and $B = \begin{bmatrix} 1 & 2 & 3 \\ 2 & 1 & 3 \\ 0 & 1 & 1 \end{bmatrix}$, Find A^T, B^T and also verify that $(A+B)^T = B^T A^T$

Q.2 If $A = \begin{bmatrix} 1 & 2 & 2 \\ 2 & 1 & -2 \\ a & 2 & b \end{bmatrix}$ is a matrix satisfying $AA^T = 9 I_3$, then find the values of a and b. (Ans: a=-2, b=-1)

Q.3 If $A=\begin{bmatrix} 2 & 3 \\ 4 & 5 \end{bmatrix}$, Prove that $A-A^T$ is a skew-symmetric matrix.

Q.4 If the matrix $A= \begin{bmatrix} 5 & 2 & x \\ y & z & -3 \\ 4 & t & -7 \end{bmatrix}$ is a symmetric matrix, find x, y, z and t.

Q.5 Let $A= \begin{bmatrix} 3 & 2 & 7 \\ 1 & 4 & 3 \\ -2 & 5 & 8 \end{bmatrix}$. Find matrices X and Y such that X+Y=A, Where X is symmetric and Y is skew symmetric matrix.

DETERMINANTS

To every square matrix A=[aij] of order n we can associate a number (real or complex) called determinant of a square matrix A.

For Example:

$A=\begin{bmatrix} a & b \\ c & d \end{bmatrix}$ then its determination denote $/A/= \begin{vmatrix} a & b \\ c & d \end{vmatrix}$

Notes * for matrix A determinate of $(|A|)$ is read as determinate of A and non-determinant.

*only square matrix have determinant.

Q 1. Evaluate the determinant, $=\begin{vmatrix} 1 & 2 & 4 \\ -1 & 3 & 0 \\ 4 & 1 & 0 \end{vmatrix}$

Sol. $\begin{vmatrix} 1 & 2 & 4 \\ -1 & 3 & 0 \\ 4 & 1 & 0 \end{vmatrix}$

Expand along R_1

$$=1\begin{vmatrix}3 & 0 \\ 1 & 0\end{vmatrix}-2\begin{vmatrix}-1 & 0 \\ 4 & 0\end{vmatrix}+1\begin{vmatrix}-1 & 3 \\ 4 & 1\end{vmatrix}$$

$$=1(0-0)-2(0-0)+4(-1-12)$$

$$=0+0+4(-13)$$

$$=-52$$

Expand along c_1

$$=1\begin{vmatrix}3 & 0 \\ 1 & 0\end{vmatrix}+1\begin{vmatrix}-2 & 4 \\ 1 & 0\end{vmatrix}+4\begin{vmatrix}-2 & 4 \\ 3 & 0\end{vmatrix}$$

$$=1(0-0)-2(0-0) +4(-1-12)$$

$$=0+0+4(-13)$$

$$=-52$$

Note:

*Value of determination can be finding separating any row one column.

*Sign of elements can be put $(-1)^{i+j}$ /Mij/

I= Row

J=column

M_{ij} =minor factor

Q.2. If $A=\begin{bmatrix}1 & 2 \\ 4 & 2\end{bmatrix}$ show that determinant of /2A/=4/A/

Sol. $A = \begin{bmatrix} 1 & 2 \\ 4 & 2 \end{bmatrix}$

L.H.S $= |2A|$

$2A = 2\begin{bmatrix} 1 & 2 \\ 4 & 2 \end{bmatrix}$

$2A = \begin{bmatrix} 2 & 4 \\ 8 & 4 \end{bmatrix}$

$|2A| = \begin{vmatrix} 2 & 4 \\ 8 & 4 \end{vmatrix}$

$/2A/ = 8 - 32$

$|2A| = -24$

R.H.S $= 4|A|$

$/A/ = \begin{bmatrix} 1 & 2 \\ 4 & 2 \end{bmatrix}$

$= 2 - 8$

$= -6$

$\therefore 4/A/ = 4(-6)$

$= -24$

Q 3. If $A = \begin{bmatrix} 1 & 0 & 1 \\ 0 & 1 & 2 \\ 0 & 0 & 4 \end{bmatrix}$ show that $/3A/ = 27/A/$

Sol. $A = \begin{bmatrix} 1 & 0 & 1 \\ 0 & 1 & 2 \\ 0 & 0 & 4 \end{bmatrix}$

L.H.S $= /3A/$

$$3A-3\begin{bmatrix} 1 & 0 & 1 \\ 0 & 1 & 2 \\ 0 & 0 & 4 \end{bmatrix}$$

$$/3A/=\begin{bmatrix} 3 & 0 & 3 \\ 0 & 3 & 6 \\ 0 & 0 & 12 \end{bmatrix}$$

Expand along R_1

$$=3\begin{bmatrix} 3 & 6 \\ 0 & 12 \end{bmatrix}-\begin{bmatrix} 0 & 6 \\ 0 & 12 \end{bmatrix}+3\begin{bmatrix} 0 & 3 \\ 0 & 0 \end{bmatrix}$$

$=-3(36-0)-(0-0)+3(0-0)$

$=103$

R.H.S$=27/A/$

$$|A|=\begin{bmatrix} 1 & 0 & 1 \\ 0 & 1 & 2 \\ 0 & 0 & 4 \end{bmatrix}$$

Expand along R_1

$$=1\begin{bmatrix} 1 & 2 \\ 0 & 4 \end{bmatrix}-\begin{bmatrix} 0 & 2 \\ 0 & 4 \end{bmatrix}+1\begin{bmatrix} 0 & 1 \\ 0 & 0 \end{bmatrix}$$

$=1(4-0)-(0-0)+1(0-0)$

$=1(4-0)$

$=4$

$\therefore 27/A/=27x4$

$=108$

PROPERTIES OF DETERMINANTS

P1*The value of determinants remind unchanged if its rows and columns are interchanged.

For eg : $= \begin{vmatrix} a1 & b1 & c1 \\ a2 & b2 & c2 \\ a3 & b3 & c3 \end{vmatrix}$

Then $= \begin{vmatrix} a1 & a2 & a3 \\ b1 & b2 & b3 \\ c1 & c2 & c3 \end{vmatrix}$

P2*If any two rows or columns of a determinant are interchanged then sign of the determinant is changed.

For eg. $= \begin{vmatrix} a1 & b1 & c1 \\ a2 & b2 & c2 \\ a3 & b3 & c3 \end{vmatrix}$

Then $= \begin{vmatrix} a1 & a2 & a3 \\ b1 & b2 & b3 \\ c1 & c2 & c3 \end{vmatrix}$

Also $\begin{vmatrix} a1 & b2 & c2 \\ a2 & b1 & c1 \\ a3 & b3 & c3 \end{vmatrix} =$

P3*If any two rows or columns of determinant are identical then the value of determinant is zero.

For eg : $\begin{vmatrix} a & a2 & a \\ b & b2 & b \\ c & c2 & c \end{vmatrix} = 0$

P4*It each element of a rows or column of a determinant

For eg: $\begin{vmatrix} a & b1 & c1 \\ a2 & b2 & c2 \\ a3 & b3 & c3 \end{vmatrix} =$

Then $\begin{vmatrix} ka1 & kb1 & kc1 \\ a2 & b2 & c2 \\ a3 & b3 & c3 \end{vmatrix} = k$

P5*If some or all elements of row or column of a determinant are expressed as sum of two or more terms then the determinant can expressed as sum of two or more determinant.

For eg: $\begin{vmatrix} a1+l & b1+m & c1+n \\ a2 & b2 & c2 \\ a3 & b3 & c3 \end{vmatrix} =$

$= \begin{vmatrix} a1 & b1 & c1 \\ a2 & b2 & c2 \\ c3 & b3 & c3 \end{vmatrix} + \begin{vmatrix} l & m & n \\ a2 & b2 & c2 \\ a3 & b3 & c3 \end{vmatrix}$

P6* If to each element of any row or column of determinant the equi-multiples of corresponding element of other moor column are added then value of determinant remind the same.

For eg. $\begin{vmatrix} a1 & b1 & c1 \\ a2 & b2 & c2 \\ a3 & b3 & c3 \end{vmatrix} =$

If apply $R_1 -> R_1 + kR_2$

$\begin{vmatrix} a1+kb1 & b1 & c1 \\ a2+kb2 & b2 & c2 \\ a3+kb3 & b3 & c3 \end{vmatrix} =$

P7* The determinant of the product of two matrix is equal to the

Q1. Without Expand $\begin{vmatrix} x+y & y+z & z+x \\ z & x & y \\ 1 & 1 & 1 \end{vmatrix} = 0$. Prove that

Sol. L.H.S $= \begin{vmatrix} x+y & y+z & z+x \\ z & x & y \\ 1 & 1 & 1 \end{vmatrix}$

Apply $R_1 \to R_1 + kR_2$

$= \begin{vmatrix} x+y+z & x+y+z & x+y+z \\ z & x & y \\ 1 & 1 & 1 \end{vmatrix}$

Taking common in R_1

$= (x+y+z) \begin{vmatrix} 1 & 1 & 1 \\ z & x & y \\ 1 & 1 & 1 \end{vmatrix}$

$= (x+y+z)(0)$

Q2. Prove that $\begin{vmatrix} b+c & a & a \\ b & c+a & b \\ c & c & a+b \end{vmatrix} = 4abc$

Sol. L.H.S $= \begin{vmatrix} b+c & a & a \\ b & c+a & b \\ c & c & a+b \end{vmatrix}$

Apply $R_1 \to R_1 - (R_2 + R_3)$

Taking common in R_1

$= 2 \begin{vmatrix} 0 & -c & -b \\ b & c+a & b \\ c & c & a+b \end{vmatrix}$

$R_1 \to R_1 + kR_2$

$$=2\begin{vmatrix} 0 & -c & -b \\ b & c+a & b \\ c & c & a+b \end{vmatrix}$$

$R_1 \to R_1 + kR_2$

$$=2\begin{vmatrix} 0 & -c & -b \\ b & a & 0 \\ c & c & a \end{vmatrix}$$

Expand along R_1

=2[0+c(ab-0)-b(0-ac)]

=2(abc+abc)

=2(2abc)

=4abc

Q3. Prove that $\begin{vmatrix} 1 & bc & a(b+c) \\ 1 & ca & b(c+a) \\ 1 & ab & c(a+b) \end{vmatrix} = 0$

Sol.   L.H.S = $\begin{vmatrix} 1 & bc & a(b+c) \\ 1 & ca & b(c+a) \\ 1 & ab & c(a+b) \end{vmatrix}$

$$=\begin{vmatrix} 1 & bc & ab+bc \\ 1 & ca & bc+ac \\ 1 & ab & ca+cb \end{vmatrix}$$

$C_3 \to c_3 + c_2$

$$=\begin{vmatrix} 1 & bc & ab+bc+ac \\ 1 & ca & ab+bc+ac \\ 1 & ab & ab+bc+ac \end{vmatrix}$$

Taking common (ab+bc+ca) in c_3

$$=(ab+bc+ca) \begin{vmatrix} 1 & bc & 1 \\ 1 & ca & 1 \\ 1 & ab & 1 \end{vmatrix}$$

$$=0$$

Q 4. Prove that $\begin{vmatrix} 2 & 7 & 65 \\ 3 & 8 & 75 \\ 5 & 9 & 86 \end{vmatrix}=0$

Sol. L.H.S$=\begin{vmatrix} 2 & 7 & 65 \\ 3 & 8 & 75 \\ 5 & 9 & 86 \end{vmatrix}=0$

Apply $C_3 -> c_3 + 9c_2$

$$=\begin{vmatrix} 2+9X7 & 7 & 65 \\ 3+9X8 & 8 & 75 \\ 5+9X9 & 9 & 86 \end{vmatrix}$$

$$=\begin{vmatrix} 65 & 7 & 65 \\ 75 & 8 & 75 \\ 86 & 9 & 86 \end{vmatrix}$$

$$=0$$

Q. Prove that $\begin{vmatrix} b+c & q+r & y+z \\ c+a & r+p & z+x \\ a+b & p+r & x+y \end{vmatrix} = 2\begin{vmatrix} a & p & x \\ b & q & y \\ c & r & z \end{vmatrix}$

Sol. L.H.S$=\begin{vmatrix} b+c & q+r & y+z \\ c+a & r+p & z+x \\ a+b & p+r & x+y \end{vmatrix}$

Apply $R_1 -> R_1 + kR_2 + R_3$

$$=\begin{vmatrix} 2(a+b+c) & 2(p+q+r) & 2(x+y+z) \\ c+a & r+p & z+x \\ a+b & p+r & x+y \end{vmatrix}$$

Apply $R_1 \to R_1 + kR_2 + R_3$

$$= \begin{vmatrix} 2(a+b+c) & 2(p+q+r) & 2(x+y+z) \\ c+a & r+p & z+x \\ a+b & p+r & x+y \end{vmatrix}$$

Taking 2 as common

$$= 2 \begin{vmatrix} a+b+c & p+q+x & x+y+z \\ c+a & r+p & z+x \\ a+b & p+r & x+y \end{vmatrix}$$

Apply $R_1 \to R_1 + kR_2 + R_3 \to R_2 - R_1$

$$= 2 \begin{vmatrix} a+b+c & p+q+r & x+y+z \\ -b & -q & -y \\ -c & -r & -z \end{vmatrix}$$

$R_1 \to R_1 + kR_2 + R_3$

$$= 2 \begin{vmatrix} a & p & x \\ -b & -q & -y \\ -c & -r & -z \end{vmatrix}$$

Taking $(-1) \times (-1)$ in R_2 and R_3

$$= 2 \begin{vmatrix} a & p & x \\ b & q & y \\ c & r & z \end{vmatrix}$$

Q. Prove that $\begin{vmatrix} -a^2 & ab & ac \\ ba & -b^2 & bc \\ ca & cb & -c^2 \end{vmatrix} = 4a^2b^2c^2$

Sol. L.H.S $= \begin{vmatrix} -a^2 & ab & ac \\ ba & -b^2 & bc \\ ca & cb & -c^2 \end{vmatrix}$

Tanking common a,b and c in R_1, R_2 and R_3 respectively

$$= abc \begin{vmatrix} -a & b & c \\ a & -b & c \\ a & b & -c \end{vmatrix}$$

Apply $R_1 \rightarrow R_1 + kR_2$

$$= abc \begin{vmatrix} -a & 0 & 2c \\ a & -b & c \\ a & b & -c \end{vmatrix}$$

Expand along R_1

$$= abc \left[0 + 0 + 2c \begin{vmatrix} a & -b \\ a & b \end{vmatrix} \right]$$

$= abc[2c(ab+ab)]$

$= abc[2c(2ab)]$

$= abc(4abc)$

$= 4a^2b^2c^2$

Q . $\begin{vmatrix} x & x^2 & yz \\ y & y^2 & zx \\ z & z^2 & xy \end{vmatrix} = (x-y)(y+z)(z-y) \ (xy+yz+zx)$. Prove that

Sol. L.H.S $= \begin{vmatrix} x & x^2 & yz \\ y & y^2 & zx \\ z & z^2 & xy \end{vmatrix}$

Apply $R_1 \rightarrow R_1 - R_2$, $R_2 \rightarrow R_2 - R_3$

$$= \begin{vmatrix} x-y & x^2-y^2 & z(x+y) \\ y-z & y^2-z^2 & x(z-y) \\ z & z^2 & xy \end{vmatrix}$$

Taking common in R_1 & R_2 (x-y), (y-z)

$$= \begin{vmatrix} 1 & x+y & -z \\ 1 & y+z & -x \\ z & z^2 & xy \end{vmatrix}$$

Apply $R_1 \to R_1 - R_2$

$$= (x-y)(y-z) \begin{vmatrix} 0 & x-y & x-z \\ 1 & y+z & -x \\ z & z^2 & xy \end{vmatrix}$$

Solving (x-y)in R_1

$$= (x-y)(x-z)(x-y) = \begin{vmatrix} 0 & 1 & 1 \\ 1 & y+z & -x \\ z & z^2 & xy \end{vmatrix}$$

Applying $c_2 \to c_2 - c_3$

$$= (x-y)(y-z)(x-z) \begin{vmatrix} 0 & 0 & 1 \\ 1 & x+y+z & -x \\ z & z^2-xy & xy \end{vmatrix}$$

Expand along R_1

$$= (x-y)(y-z)(x-z) \begin{vmatrix} 1 & x+y+z \\ z & z^2-xy \end{vmatrix}$$

$$= (x-y)(y-z)(x-z)[z^2xy\text{-}xz\text{-}yz\text{-}z^2]$$

Q. Prove that $\begin{vmatrix} 1 & 1+p & 1+p+q \\ 2 & 3+2p & 4+3p+2q \\ 3 & 6+3p & 10+6p+3q \end{vmatrix} = 1$

Sol. L.H.S $= \begin{vmatrix} 1 & 1+p & 1+p+q \\ 2 & 3+2p & 4+3p+2q \\ 3 & 6+3p & 10+6p+3q \end{vmatrix}$

We know than if any row or column are sum of our or two or more elements then the determinant can be expressed as sums of two or more determinant.

$= \begin{vmatrix} 1 & 1 & 1+p+q \\ 2 & 3 & 4+3p+2q \\ 3 & 6 & 10+6p+3q \end{vmatrix} + \begin{vmatrix} 1 & p & 1+p+q \\ 2 & 2p & 4+3p+2q \\ 3 & 3p & 10+6p+3q \end{vmatrix}$

$= \begin{vmatrix} 1 & 1 & 1 \\ 2 & 3 & 4 \\ 3 & 6 & 10 \end{vmatrix} + \begin{vmatrix} 1 & 1 & p \\ 2 & 3 & 3p \\ 3 & 6 & 6p \end{vmatrix} + \begin{vmatrix} 1 & 1 & q \\ 2 & 3 & 2q \\ 3 & 6 & 3q \end{vmatrix} + p\begin{vmatrix} 1 & 1 & 1 \\ 2 & 2 & 2 \\ 3 & 3 & 3 \end{vmatrix}$

$C_1 \text{->} c_1 - c_2, C_2 \rightarrow C_2 - C_3$

$= \begin{vmatrix} 0 & 0 & 1 \\ -1 & -1 & 4 \\ -3 & -4 & 10 \end{vmatrix} + p\begin{vmatrix} 1 & 1 & 1 \\ 2 & 3 & 3 \\ 3 & 6 & 6 \end{vmatrix} + q\begin{vmatrix} 1 & 1 & 1 \\ 2 & 3 & 2 \\ 3 & 6 & 3 \end{vmatrix} + p(0)$

Expand along R_1

$= \begin{vmatrix} -1 & -1 \\ -3 & -4 \end{vmatrix} + p(0) + q(0) + p(0)$

$= 1$

Applications of determinants

*Area of triangle ABC with vertices $A(x_1, y_1)$, $B(x_2, y_2)$ and $c(x_3, y_3)$ is defined as follows

$$\frac{1}{2}\begin{vmatrix} x_1 & y_1 & 1 \\ x_2 & y_2 & 1 \\ x_3 & y_3 & 1 \end{vmatrix}$$

Q. Find area of triangle with vertices A (2,3), B(-1,4), c(5,3)

$$=\frac{1}{2}\begin{vmatrix} x & y & 1 \\ x_1 & y_1 & 1 \\ x_2 & y_3 & 1 \end{vmatrix}$$

Sol. Area (ABC)$=\frac{1}{2}\begin{vmatrix} 2 & 3 & 1 \\ -1 & 4 & 1 \\ 5 & 3 & 1 \end{vmatrix}$

Apply $R_1 \rightarrow R_1 - R_2$, $R_2 \rightarrow R_2 - R_3$

$$=\frac{1}{2}\begin{vmatrix} 3 & -1 & 0 \\ -6 & 1 & 0 \\ 5 & 3 & 1 \end{vmatrix}$$

Expands along C_3

$$=\frac{1}{2}\begin{vmatrix} 3 & -1 \\ -6 & 1 \end{vmatrix}$$

$$=\frac{1}{2}(3\text{-}6) =\frac{1}{2}(\text{-}3) =\frac{-3}{2}$$

$$=\frac{3}{2}\text{squre unit}$$

Q. prove that $\begin{vmatrix} 1+a & 1 & 1 \\ 1 & 1+b & 1 \\ 1 & 1 & 1+c \end{vmatrix}= abc\ (1\tfrac{1}{a}+\tfrac{1}{b}+\tfrac{1}{c})$
$= abc+bc+ca+ab$

Sol. L.H.S$=\begin{vmatrix} 1+a & 1 & 1 \\ 1 & 1+b & 1 \\ 1 & 1 & 1+c \end{vmatrix}$

$$= abc\ (1\tfrac{1}{a}+\tfrac{1}{b}+\tfrac{1}{c})\begin{vmatrix} 1 & 1 & 1 \\ \tfrac{1}{b} & \tfrac{1}{b}+1 & \tfrac{1}{b} \\ \tfrac{1}{c} & \tfrac{1}{c} & \tfrac{1}{c}+1 \end{vmatrix}$$

Apply

$$= abc\ (1\tfrac{1}{a}+\tfrac{1}{b}+\tfrac{1}{c})\begin{vmatrix} 0 & 0 & - \\ -1 & 1 & \tfrac{1}{b} \\ 0 & -1 & \tfrac{1}{c}+1 \end{vmatrix}$$

Expand along

$$= abc\ (1\tfrac{1}{a}+\tfrac{1}{b}+\tfrac{1}{c})\begin{vmatrix} -1 & 1 \\ 0 & -1 \end{vmatrix}$$

$$= abc\ (1\tfrac{1}{a}+\tfrac{1}{b}+\tfrac{1}{c})[(-1)\text{-}(0)]$$

$$= abc\ (1+\tfrac{1}{a}+\tfrac{1}{b}+\tfrac{1}{c})(1)$$

$$= abc+ab+bc+ca$$

PROBLEMS FOR PRACTICE

By using properties of determinants prove the following.

Q.1 $\begin{vmatrix} -a^2 & ab & ac \\ ba & -b^2 & bc \\ ca & bc & -c^2 \end{vmatrix} = 4a^2b^2c^2$

Q.2 $\begin{vmatrix} a-b-c & 2a & 2a \\ 2b & b-c-a & 2b \\ 2c & 2c & c-a-b \end{vmatrix} = (a+b+c)^3$

Q.3 $\begin{vmatrix} x+y+2z & x & y \\ z & y+z+2x & y \\ z & x & z+x+2y \end{vmatrix} =$

$2(x+y+z)^3$

Q.4 $\begin{vmatrix} 1 & x & x^2 \\ x^2 & 1 & x \\ x & x^2 & 1 \end{vmatrix} = (1\text{-}x^3)^2$

Q.5 $\begin{vmatrix} a^2 & bc & ac+c^2 \\ a^2+ab & b^2 & ac \\ ab & b^2+bc & c^2 \end{vmatrix} = 4a^2b^2c^2$

Q.6 $\begin{vmatrix} a^2 & bc & ac+c^2 \\ a^2+ab & b^2 & ac \\ ab & b^2+bc & c^2 \end{vmatrix} = 4a^2b^2c^2$

Q.7 $\begin{vmatrix} a & b & c \\ a^2 & b^2 & c^2 \\ b+c & c+a & a+b \end{vmatrix} = (b\text{-}c)\,(c\text{-}a)(a\text{-}b)(a+b+c)$

Q.8 $\begin{vmatrix} 1 & 1 & 1 \\ \alpha^2 & \beta^2 & \gamma^2 \\ \alpha^3 & \beta^3 & \gamma^3 \end{vmatrix} = (\alpha - \beta)(\beta - \gamma)(\gamma - \alpha)(\alpha\beta + $

$\beta\gamma + \gamma\alpha)$

Application of determinants can be used to solve a system of linear equations. We shall study the following theorem.

Consider the following equations

$a_1x+b_1y=c_1$

$a_2x+b_2y = c_2$

$\Delta = \begin{vmatrix} a_1 & b_1 \\ a_2 & b_2 \end{vmatrix}$ $\quad \Delta x = \begin{vmatrix} a_1 & b_1 \\ a_2 & b_2 \end{vmatrix}$ and $\Delta y = \begin{vmatrix} a_1 & b_1 \\ a_2 & b_2 \end{vmatrix}$

We can classify the solution of the given equation in three different solution on the basis of value of Δ Δx and Δy

   i.   If $\Delta \neq 0$, then the system of equation has unique solution.

   ii.   If $\Delta = 0$, and, $\Delta x \neq 0$, $\Delta y \neq 0$, then the equation has No solution system.

   iii. If $\Delta = 0$, $\Delta x = 0$ and $\Delta y = 0$, then the equation has infinite number of solutions.

Similarly we can apply in three variable equations

$a_1x+b_1y+c_1z=d_1$, $a_2x+b_2y+c_2z=d_2$ and $a_3x+b_3y+c_3z=d_3$

Here, $\Delta = \begin{vmatrix} a_1 & b_1 & c_1 \\ a_2 & b_2 & c_2 \\ a_3 & b_3 & c_3 \end{vmatrix}$, $\Delta x = \begin{vmatrix} d_1 & b_1 & c_1 \\ d_2 & b_2 & c_2 \\ d_3 & b_3 & c_3 \end{vmatrix}$,

$\Delta y = \begin{vmatrix} a_1 & d_1 & c_1 \\ a_2 & d_2 & c_2 \\ a_3 & d_3 & c_3 \end{vmatrix}$ and , $\Delta z = \begin{vmatrix} a_1 & b_1 & d_1 \\ a_2 & b_2 & d_2 \\ a_3 & b_3 & d_3 \end{vmatrix}$

If the equation has unique solution system then we can find the x,y and z by the following equations

$X = \frac{\Delta x}{\Delta}$, $y = = \frac{\Delta y}{\Delta}$, $z = \frac{\Delta z}{\Delta}$ (This rule is called **Cramer's RULE** for solution of a system of linear Equation.

Q.1 Solve the following system of equation by means of determinants.

$X+2y+3z=6$, $2x+4y+z=7$, and $3x+2y+9z=14$

Sol. From the above equation

$$\Delta = \begin{vmatrix} a_1 & b_1 & c_1 \\ a_2 & b_2 & c_2 \\ a_3 & b_3 & c_3 \end{vmatrix} \Rightarrow \begin{vmatrix} 1 & 2 & 3 \\ 2 & 4 & 1 \\ 3 & 2 & 9 \end{vmatrix} = -20 \neq 0$$

So, the system has unique solution

$$\text{Now, } \Delta x = \begin{vmatrix} d_1 & b_1 & c_1 \\ d_2 & b_2 & c_2 \\ d_3 & b_3 & c_3 \end{vmatrix} = \begin{vmatrix} 6 & 2 & 3 \\ 7 & 4 & 1 \\ 14 & 2 & 9 \end{vmatrix} = -20$$

$$\Delta y = \begin{vmatrix} a_1 & d_1 & c_1 \\ a_2 & d_2 & c_2 \\ a_3 & d_3 & c_3 \end{vmatrix} = \begin{vmatrix} 1 & 6 & 3 \\ 2 & 7 & 1 \\ 3 & 14 & 9 \end{vmatrix} = -20$$

$$\Delta z = \begin{vmatrix} a_1 & b_1 & d_1 \\ a_2 & b_2 & d_2 \\ a_3 & b_3 & d_3 \end{vmatrix} = \begin{vmatrix} 1 & 2 & 6 \\ 2 & 4 & 7 \\ 3 & 2 & 14 \end{vmatrix} = -20$$

$$X = \frac{\Delta x}{\Delta} = 1 \ , \ y = = \frac{\Delta y}{\Delta} = 1 \ , z = \frac{\Delta z}{\Delta} = 1$$

PROBLEMS FOR PRACTICE

Solve the following equation by using Cramer's rule (Determinant method)

Q.1 $3x+5y-7z=13$, $4x+y-12z=6$ and $2x+9y-3z=20$
     Ans: x=1, y=2, z=0

Q.2 $x+y+z=9$, $2x+5y+7z=52$, and $2x+y+z=0$
     Ans; x=1, y=3, z=5

Q.3 $x+y+z=1$, $2x+3y+z=4$,   and $4x+9y+z=16$
     Ans: x=-3, y=3, z=1

ADJOINT AND INVERSE OF A MATRIX

Definition: Let A= $\lceil a_{ij} \rceil$ be a square matrix of order n and C_{IJ} be cofactor of a_{ij} in A. Then the transpose of the matrix of cofactors of elements of A is called the adjoint of A and is denoted by adjA.

Q.Find the adjont of matrix A= $\lceil a_{ij} \rceil = \begin{bmatrix} 1 & 1 & 1 \\ 2 & 1 & -3 \\ -1 & 2 & 3 \end{bmatrix}$

Sol. Let C_{IJ}be factors of a_{ij}in A. Then, the cofactors of elements of A are given by

$C_{11}=\begin{vmatrix} 1 & -3 \\ 5 & -2 \end{vmatrix}=9 \qquad C_{12}=-\begin{vmatrix} 2 & -3 \\ -1 & 3 \end{vmatrix}=\text{-3},$

$C_{13}=\begin{vmatrix} 2 & 1 \\ -1 & 2 \end{vmatrix}=5$

$C_{21}=-\begin{vmatrix} 1 & 1 \\ 2 & 3 \end{vmatrix}=\text{-1}, \qquad C_{22}=\begin{vmatrix} 1 & 1 \\ -1 & 3 \end{vmatrix}=4,$

$C_{23}=-\begin{vmatrix} 1 & 1 \\ -1 & 2 \end{vmatrix}=\text{-3}$

$C_{31}=\begin{vmatrix} 1 & 1 \\ 1 & -3 \end{vmatrix}=\text{-4}, \qquad C_{32}=-\begin{vmatrix} 1 & 1 \\ 2 & -3 \end{vmatrix}=5,$

$C_{33}=\begin{vmatrix} 1 & 1 \\ 2 & 1 \end{vmatrix}=\text{-1}$

$\therefore \text{ adjA}=\begin{bmatrix} 9 & -3 & 5 \\ -1 & 4 & -3 \\ -4 & 5 & -1 \end{bmatrix}^{T}=\begin{bmatrix} 9 & -1 & -4 \\ -3 & 4 & 5 \\ 5 & -3 & -1 \end{bmatrix}$

INVERSE OF A MATRIX

Definition: A square matrix of order n is invertible if there exists a square matrix B of the same order such that

AB=BA=I

In such cases, we say that the inverse of A is B and we write, $A^{-1}=B$

We have a formula for finding the inverse of a Non-singular square matrix. The inverse of A given by

$A^{-1}=\frac{1}{|A|}.adjA$

Q. Find the inverse of A if $A=\begin{bmatrix} 2 & -1 \\ 3 & 4 \end{bmatrix}$

Sol. Let $A=\begin{bmatrix} 2 & -1 \\ 3 & 4 \end{bmatrix}$, then

$|A| =\begin{vmatrix} 2 & -1 \\ 3 & 4 \end{vmatrix}= 8+3=11\neq 0$

So, A is a non- singular matrix and therefore it is invertible. Now, we need to find adjA

$\therefore C_{11}=4, \quad C_{12}=-3, \quad C_{21}=1, \quad C_{22}=2$ (Already we discussed about cofactors)

$\therefore adjA=\begin{bmatrix} 4 & -3 \\ 1 & 2 \end{bmatrix}^T=\begin{bmatrix} 4 & 1 \\ -3 & 2 \end{bmatrix}$

Hence, $A^{-1}=\frac{1}{|A|}adjA =\frac{1}{11}\begin{bmatrix} 4 & 1 \\ -3 & 2 \end{bmatrix}$

$=\begin{bmatrix} 4/11 & 1/11 \\ -3/11 & 2/11 \end{bmatrix}$

Q.Find the inverse of the matrix $A =\begin{bmatrix} 8 & 4 & 2 \\ 2 & 9 & 4 \\ 1 & 2 & 8 \end{bmatrix}$

Sol. We have,

$$|A| = \begin{vmatrix} 8 & 4 & 2 \\ 2 & 9 & 4 \\ 1 & 2 & 3 \end{vmatrix} = 8(72-8)-4(16-4) +2(4-9) =454\neq 0$$

So, A is non-singular matrix and therefore it is invertible.

The cofactors of A are

$C_{11}=64, \quad C_{12}=-12, \quad C_{13}=-5$

$C_{21}=-28, \quad C_{22}=62, \quad C_{23}=-12$

$C_{31}=-2, \quad C_{32}=-28, \quad C_{33}=64$

$$AdjA = \begin{bmatrix} 64 & -12 & -5 \\ -28 & 62 & -12 \\ -2 & -28 & 64 \end{bmatrix}^T = \begin{bmatrix} 64 & -28 & -2 \\ -12 & 62 & -28 \\ -5 & -12 & 64 \end{bmatrix}$$

Hence, $A^{-1} = A^{-1} = \frac{1}{|A|}adjA$

$$= \frac{1}{454} \begin{bmatrix} 64 & -28 & -2 \\ -12 & 62 & -28 \\ -5 & -12 & 64 \end{bmatrix}$$

PROBLEMS FOR PRACTICE

Q.1 Find adjoint matrix of A, if $A = \begin{bmatrix} 3 & 1 & 2 \\ 2 & 2 & 5 \\ 4 & 1 & 0 \end{bmatrix}$

Ans: $adjA = \begin{bmatrix} 3 & 1 & 2 \\ 2 & 2 & 5 \\ 4 & 1 & 0 \end{bmatrix}$

Q.2 Find the inverse of A, if $A = \begin{bmatrix} 2 & 3 & 4 \\ 3 & 2 & 1 \\ 1 & 1 & -2 \end{bmatrix}$ Ans:

$$A^{-1} = \frac{1}{15} \begin{bmatrix} -5 & 10 & -5 \\ 7 & -8 & 10 \\ 1 & 1 & -5 \end{bmatrix}$$

Matrix method for the solution of a Homogeneous system of simultaneous linear equations

To obtain the system of equations we consider the equations

$a_1x + b_1y + c_1z = d_1, \quad a_2x + b_2y + c_2z = d_2$ and $a_3x + b_3y + c_3z = d_3$

We can rewrite in matrix method as follows

$$\begin{bmatrix} a_1 & b_1 & c_1 \\ a_2 & b_2 & c_2 \\ a_3 & b_3 & c_3 \end{bmatrix} \begin{bmatrix} x \\ y \\ z \end{bmatrix} = \begin{bmatrix} d_1 \\ d_2 \\ d_3 \end{bmatrix}$$

$$A X = B$$

From the value of $|A|$ we can classify the solution as follows.

(i) If $|A| \neq 0$, then the system of equations is consistent with unique solution and we can find the value of X by using the identity $X = A^{-1}B$.

(ii) If $|A| = 0$, then the given system of equations is either inconsistent or it has infinitely many solution. It can be confirmed by the following workings

a) If (adjA) B≠0, then the system will be inconsistent or no solution system.

b) If (adjA) B=0, then the system will be consistent with infinitely many solutions.

Q.1 Use matrix to solve the following equations

5x-7y=2, 7x-5y=3

Sol. The given equations are

5x-7y=2--------(i)

7x-5y=3-------(ii)

We can rewrite in matrix method

$$\begin{bmatrix} 5 & -7 \\ 7 & -5 \end{bmatrix} \begin{bmatrix} x \\ y \end{bmatrix} = \begin{bmatrix} 2 \\ 3 \end{bmatrix}$$

Let A X= B, where

$$A = \begin{bmatrix} 5 & -7 \\ 7 & -5 \end{bmatrix} \quad X = \begin{bmatrix} x \\ y \end{bmatrix} \quad B = \begin{bmatrix} 2 \\ 3 \end{bmatrix}$$

$Now|A| = \begin{vmatrix} 5 & -7 \\ 7 & -5 \end{vmatrix} = -25+49 = 24 \neq 0$

So, the system has unique solution system

$X = A^{-1}B$

To find A^{-1} ,we need to find cofactors

$C_{11} = -5$, $C_{12} = -7, C_{21} = 7, C_{22} = 5$

$Adj.A = \begin{bmatrix} 5 & -7 \\ 7 & 5 \end{bmatrix}^T = \begin{bmatrix} -5 & 7 \\ -7 & 5 \end{bmatrix}$

164 | *Mathematics for Commerce*

$$A^{-1}=\frac{1}{|A|}adj.A =\frac{1}{24}\begin{bmatrix}-5 & 7\\-7 & 5\end{bmatrix}\begin{bmatrix}2\\3\end{bmatrix}$$

$$=\frac{1}{24}\begin{bmatrix}-10+21\\-14+15\end{bmatrix}=\begin{bmatrix}11/24\\1/24\end{bmatrix}$$

$$\begin{bmatrix}x\\y\end{bmatrix}=\begin{bmatrix}\frac{11}{24}\\\frac{1}{24}\end{bmatrix}$$

X=11/24, Y=1/24

Q.2 Solve the following system of equation, using matrix method:

x+2y+z=7,x+3z=11,2x-3y=1

Sol. The given equation can be represents in matrix as follows

$$\begin{bmatrix}1 & 2 & 1\\1 & 0 & 3\\2 & -3 & 0\end{bmatrix}\begin{bmatrix}x\\y\\z\end{bmatrix}=\begin{bmatrix}7\\11\\1\end{bmatrix}$$

Let AX=B⇒X=A⁻¹B

Where $A=\begin{bmatrix}1 & 2 & 1\\1 & 0 & 3\\2 & -3 & 0\end{bmatrix}$ $X=\begin{bmatrix}x\\y\\z\end{bmatrix}$ and $B=\begin{bmatrix}7\\11\\1\end{bmatrix}$

Now, $|A|=\begin{vmatrix}1 & 2 & 1\\1 & 0 & 3\\2 & -3 & 0\end{vmatrix}=1(0+9)-2(0-6)+1(-3-0)=18\neq 0$

So, the system has unique solution system,

To find A⁻¹,we need to find cofactors

$C_{11}=9$ $C_{12}=9$, $C_{13}=-3$

$C_{21}=-3$ $C_{22}=-2$, $C_{23}=7$

$C_{31}=6$ $C_{32}=-2$, $C_{33}=-2$

$$\therefore \text{adj}A=\begin{bmatrix} 9 & 6 & -3 \\ -3 & -2 & 7 \\ 6 & -2 & -2 \end{bmatrix}^T=\begin{bmatrix} 9 & -3 & 6 \\ 6 & -2 & -2 \\ -3 & 7 & -2 \end{bmatrix}$$

$$A^{-1}=\frac{1}{|A|}(\text{adj}.A)$$

$$=\frac{1}{18}\begin{bmatrix} 9 & -3 & 6 \\ 6 & -2 & -2 \\ -3 & 7 & -2 \end{bmatrix}$$

Now $=A^{-1}B$

$$X=\frac{1}{18}\begin{bmatrix} 9 & -3 & 6 \\ 6 & -2 & -2 \\ -3 & 7 & -2 \end{bmatrix}\begin{bmatrix} 7 \\ 11 \\ 1 \end{bmatrix}$$

$$=\frac{1}{18}\begin{bmatrix} 63-33+6 \\ 42-22-2 \\ -21+77-2 \end{bmatrix}$$

$$=\frac{1}{18}\begin{bmatrix} 36 \\ 18 \\ 54 \end{bmatrix}$$

$$\begin{bmatrix} x \\ y \\ z \end{bmatrix}=\frac{1}{18}\begin{bmatrix} 36 \\ 18 \\ 54 \end{bmatrix}=\begin{bmatrix} 2 \\ 1 \\ 3 \end{bmatrix} \Rightarrow x=2, y=1, z=3$$

Q.The sum of three numbers is 6. If we multiply by third number by 2 and add the first number to the result ,we get 7.By adding second and third numbers to three times the first number, we get 12.Using matrices find the number.

Sol. Let the three numbers be x,y and z respectively. Then,

$$x+y+z=6, \quad x+2z=7 \quad , 3x+y+z=12$$

The above equations can be rewrite in matrices as follows

$$\begin{bmatrix} 1 & 1 & 1 \\ 1 & 0 & 2 \\ 3 & 1 & 1 \end{bmatrix} \begin{bmatrix} x \\ y \\ z \end{bmatrix} = \begin{bmatrix} 6 \\ 7 \\ 12 \end{bmatrix}$$

Let A X=B

Where , $A=\begin{bmatrix} 1 & 1 & 1 \\ 1 & 0 & 2 \\ 3 & 1 & 1 \end{bmatrix}$ $X=\begin{bmatrix} x \\ y \\ z \end{bmatrix}$ $B=\begin{bmatrix} 6 \\ 7 \\ 12 \end{bmatrix}$

Now, $|A|=\begin{vmatrix} 1 & 1 & 1 \\ 1 & 0 & 2 \\ 3 & 1 & 1 \end{vmatrix}=1(0-2)-1(1-6)+1(1-0)=4\neq 0$

So, the system of solution has unique solution system

$X=A^{-1}B$

To find A^{-1} we need cofactors and the cofactors are

$C_{11}=-2, \ C_{12}=5, \ C_{13}=1$

$C_{21}=0, \ C_{22}=-2, C_{23}=2$

$C_{31}=2, \ C_{32}=-1$ and $C_{33}=-1$

$$AdjA=\begin{bmatrix} -2 & 5 & 1 \\ 0 & -2 & 2 \\ 2 & -1 & -1 \end{bmatrix}^{T} \Rightarrow \begin{bmatrix} -2 & 0 & 2 \\ 5 & -2 & -1 \\ 1 & 2 & -1 \end{bmatrix}$$

$$A^{-1}=\frac{1}{|A|}adjA=\frac{1}{4}\begin{bmatrix}-2 & 0 & 2\\ 5 & -2 & -1\\ 1 & 2 & -1\end{bmatrix}$$

Now, $X=A^{-1}B$

$$\Rightarrow \frac{1}{4}\begin{bmatrix}-2 & 0 & 2\\ 5 & -2 & -1\\ 1 & 2 & -1\end{bmatrix}\begin{bmatrix}6\\ 7\\ 12\end{bmatrix}$$

$$=\frac{1}{4}\begin{bmatrix}-12+0+24\\ 30-14-12\\ 6+14-12\end{bmatrix}$$

$$\begin{bmatrix}X\\ Y\\ Z\end{bmatrix}=\frac{1}{4}\begin{bmatrix}12\\ 4\\ 8\end{bmatrix}=\begin{bmatrix}3\\ 1\\ 2\end{bmatrix}$$

Hence, the numbers are 3,1 and 2.

PROBLEMS FOR PRACTICE

Solve the following equations by using Matrix method

Q.1 5x+7y+2=0, and 4x+6y+3=0
(Ans: x=9/2, y=-7/2)

Q.2 3x+7y=4, and x+2y=-1
(Ans: x=-15, y= 7)

Q.3 x+y-z=3, 2x+3y+z=10 and 3x-y-7z=1
(Ans: x=3, y=1,z=1)

Q.4 x+y+z=3, 2x-y+z=-1 and 2x+y-3z=-9
(Ans: x=-8/7,y=10/7,z=19/7)

Q.5 3x+4y+2z=8, 2y-3z=3 and x-2y+6z=-2
(Ans:x=-2, y=3, z=1)

Q.6 An amount of ₹ 10,000 is put onto three investments at the rate of10, 12and 15% per Annum. The combined income is ₹1310 and the combined income of first and second is ₹190 short of the income from third. Find the investment in each using Matrices method. (Ans: ₹2, 000, ₹ 3,000 and ₹ 5,000)

Q.7 A company produces three products every day. Their production on a certain day is 45 tons. It is found that the production of third product exceeds the production of first product by 8 tons while the total production of first and third product is twice the production of second product. Determine the production level of each product using matrix method. (Ans:11 tons, 15tons and 19tons)

DIFFERENTIATION

Most of the economic decisions are based on the solution to the following basic assumptions

"Whether a particular line of action would comparatively add more to our benefits than the efforts spent for pursuing it".

Such a question is of vital importance for making final decision or taking any project or solving any economic problem. In economics, this question is the core of marginal Analysis-which analyses the change in the overall performance due to marginal change consciously made the value of the variable in question. So the marginal analysis is closely related to mathematical technique known as differential calculus.

The analysis is business and economics is frequently concerned with change, calculus is an extremely valuable tool in solving problems in these fields. Marginal analysis is one of the important applications of calculus in business and economics.

Limits: Limit is a function to find the slight changes in the dependent variable with respect to the change in independent variable. The following are the formulas applicable to find the limit calculations.

$$\lim_{x\to a}(f(x) \pm g(x)) = \lim_{x\to a}f(x) \pm \lim_{x\to a}g(x)$$

$$\lim_{x\to a}(f(x).g(x)) = \lim_{x\to a}f(x).\lim_{x\to a}g(x)$$

$$\lim_{x\to a}(f(x)/g(x)) = \lim_{x\to a}f(x)/\lim_{x\to a}g(x)$$

$$\lim_{x\to a}\frac{x^n-a^n}{x-a} = na^{n-1}$$

$$\lim_{x\to 0}\frac{e^x-1}{x} = 1$$

$$\lim_{x\to 0}\frac{1}{x} = \infty$$

The value of limit is applicable is directly, when the given limit function not provide '$\frac{0}{0}$' form or '$\frac{A}{0}$' form

A-Any real value.

Q.1 Evaluate $\lim_{x\to 2}\dfrac{x^3-8}{x-2}$

Sol. $\lim_{x\to 2}\dfrac{x^3-8}{x-2}$ (if we put x=0, we get $\frac{0}{0}$)

So, we apply a method to get exist the limit

$$= \lim_{x\to 2}\frac{x-2)(x^2+2x+4)}{x-2}$$ (factor method)

$$= \lim_{x\to 2}(x^2 + 2x + 4) = 4+4+4 = 12$$

Q.2 $\lim_{x\to 3}\dfrac{x^2-9}{3-3}$

Sol. $\lim_{x\to 3}\dfrac{x^2-9}{3-3}$

$$= \frac{lim}{x \to 3} \frac{x^2 - 3^2}{x - 3}$$

$$= 2(3)^{2-1} \quad (\frac{lim}{x \to a} \frac{x^n - a^n}{x - a} = na^{n-1})$$

$$= 2 \times 3$$

$$= 6$$

Differentiation

Let us know what does mean differentiation, when a function is dependent any variable, and then change of dependent value must change the value of independent variable. The rate of change of this relation is known as differentiation and it is defined $\frac{dy}{dx} = \frac{lim}{\Delta x \to 0} \frac{f(x+\Delta x) - f(x)}{\Delta x}$, is instantaneous rate of change of y with respect to x (w.r.t.x)

For example, $f(x) = x^2$ we can find its differentiation by using the limit method it also called as first principal method.

$F(x) = x^2$ i.e. $y = x^2$ .... (i)

Δy be the change in y with respect to change of Δx in x

$Y + \Delta y = (x + \Delta x)^2$ ......(ii)

Subtract (i) from (ii)

$\Delta y = (x + \Delta x)^2 - x^2$

Dividing Δx for both the sides we get

$$\frac{\Delta y}{\Delta x} = \frac{(x+\Delta x)2-x^2}{\Delta x}$$

$$\lim_{\Delta x \to 0} \frac{\Delta y}{\Delta x} = \lim_{\Delta x \to 0} \frac{(x+\Delta x)2-x^2}{\Delta x}$$

$$\frac{dy}{dx} = \frac{x^2+2x\Delta x+\Delta x^2-x^2}{\Delta x}$$

$$= \lim_{\Delta x \to 0} \frac{\Delta x(2x+\Delta x)}{\Delta x}$$

$$\frac{dy}{dx} = 2x$$

Formulas

Rule: 1. The derivative of a constant is Zero. $\frac{d}{dx}(c)=0$

Rule: 2. The derivative of x^n is $\frac{d}{dx}(x^n) = nx^{n-1}$

Rule: 3. The derivative of u^n is $\frac{d}{dx}(u^n) = nu^{n-1}\frac{d(U)}{dx}$

Rule: 4. Sum or difference of derivative of the function $\frac{d}{dx}(U \pm V) = \frac{d}{dx}(U) \pm \frac{d}{dx}(V)$

Rule: 5 Product of two different function $\frac{d}{dx}(UV) = U\frac{d}{dx}(v) + V\frac{d}{dx}(U)$

Rule: 6 Division of two different function $\frac{d}{dx}\left(\frac{U}{V}\right) = \frac{V\frac{dU}{dx} - u\frac{dv}{dx}}{V^2}$

Rule: 7 Differentiation of log function $\frac{d\,(logx)}{dx} = \frac{1}{x}$

Rule: 8 Differentiation of exponential function

$$\frac{d(e)^x}{dx} = e^x$$

Rule: 9 Chain rule when function defined in third variable say 't' then $\frac{dy}{dx} = \frac{dx}{dt} \times \frac{dt}{dy}$

Rule: 10 Successive differentiation of a function Y is known as first order, second order, third order...

Denoted as $\frac{dy}{dx}$, d^2y/dx^2, d^3y/dx^3 ...... respectively.

We can clear the concept of differentiation by the following illustration.

Q.1. Differentiate x^9 w.r.t.x

Sol. Let $y = x^9$

Differentiate w.r.t.x

$$\frac{dy}{dx} = \frac{d(x)9}{dx} = 9x^{9-1} = 9x^8 \quad (\text{using} \frac{d}{dx}(x^n) = nx^{n-1})$$

Q.2. Differentiate $(3x+7)^7$ w.r.t.x

Sol. Let $Y = (3x+7)^7$

$$\frac{dy}{dx} = 7(3x+7)^{7-1} \frac{d(3x+7)}{dx} \quad (\text{Using} \frac{d}{dx}(u^n) = nu^{n-1} \frac{d(U)}{dx})$$

$$= 7(3x+7)^6 . 3$$

$$= 21(3x+7)^6$$

Q.3 Differentiate: $x^8 + 15x^7 - 5x^4 - 3x^3 - 12x^2 - 7$

Sol. Let $Y = x^8 + 15x^7 - 5x^4 - 3x^3 - 12x^2 - 7$

Diff.w.r.t.x

$$\frac{dy}{dx} = \frac{d}{dx}(x^8+15x^7-5x^4-3x^3-12x^2-7)$$

$$= \frac{d}{dx}(x^8) + \frac{d}{dx}(15x^7) - \frac{d}{dx}(5x^4) - \frac{d}{dx}3X^3 - \frac{d}{dx}(12X^2) - \frac{d}{dx}(7)$$

$$(\text{Using } \frac{d}{dx}(U\pm V) = \frac{d}{dx}(U) \pm \frac{d}{dx}(V)$$

$$= 8X^7 + 15.7\ X^6 - 5.4\ X^3 - 3.3X^2 - 12.2X - 0$$

$$\frac{dy}{dx} = 8X^7 + 135\ X^6 - 20\ X^3 - 9X^2 - 24X \text{Type equation here.}$$

Q.4 Differentiate: $(2x^2-5)(x^3+2x)$

Sol. Let $y = (2x^2-5)(x^3+2x)$

Diff.w.r.t.x

$$\frac{dy}{dx} = \frac{d}{dx}(2x^2-5)(x^3+2x)$$

$$= (2x^2-5)\frac{d}{dx}(x^3+2x) + (x^3+2x)\frac{d}{dx}(2x^2-5)$$

$$= (2x^2-5)(3x^2+2) + (x^3+2x)\ 4x$$

$$= 6x^4+4x^2-15x^2-10+4x^4+8x^2$$

$$\frac{dy}{dx} = 10x^4-3x^2-10$$

Q.5 Differentiate w.r.t.x $(x^2+1)/(x^3-5)$

Sol. Let $y = (x^2+1)/(x^3-5)$

Diff.w.r.t.x

$$\frac{dy}{dx} = \frac{d}{dx}(x^2+1)/(x^3-5)$$

Using the rule: $\frac{d}{dx}(\frac{U}{V}) = \frac{V\frac{dU}{dx} - u\frac{dv}{dx}}{V^2}$

Using the formula

$$= (x^3-5)\frac{d}{dx}(x^2+1) - (x^2+1)\frac{d}{dx}(x^3-5)/(x^3-5)^2$$

$$=(x^3-5)(2x) - (x^2+1)3x^2 /(x^3-5)^2$$

$$= 2x^3-10x -3x^4+3x^2/(x^3-5)^2$$

$$= -3x^4+2x^3+3x^2-10x/(x^3-5)^2$$

Q.6 Differentiate $\log(x^2+3x+4)$

Sol. Let Y= $\log(x^2+3x+4)$

Diff.w.r.t.x

$$\frac{dy}{dx} = \frac{d}{dx}(\log(x^2+3x+4))$$

$$=\frac{1}{(x2+3x+4))}\frac{d}{dx}((x^2+3x+4)$$

(using function $\frac{d\ (logx)}{dx} = \frac{1}{x}$)

$$=\frac{1}{(x2+3x+4))}(2x + 3)$$

$$= 2x+3/\log(x^2+3x+4)$$

Q.7 e^{2x+3}

Sol. Let y $=e^{2x+3}$

Diff.w.r.t.x

$$\frac{dy}{dx} = \frac{d}{dx}(e^{2x+3})$$

$$= e^{2x+3} \frac{d}{dx}(2x+3) \quad \left(\text{Using } \frac{d(e)^x}{dx} = e^x\right)$$

$$\frac{dy}{dx} = e^{2x+3}(2)$$

Q.8 Find $\frac{dy}{dx}$ when y= $5z^5+10z$, x=$2z^3-z^2$

Sol. Here the function x and y are defined in terms of third variable is called as parametric form

So, we have to use the chain rule to find $\frac{dy}{dx}$

i.e. $\frac{dy}{dx} = \frac{dy}{dz} \times \frac{dz}{dx}$

Now, y =$5z^5+10z$

Diff.w.r.t.z

$$\frac{dy}{dz} = \frac{d}{dz}(5z^5+10z)$$

$$\frac{dy}{dz} = 25z^4+10 \text{-------(i)}$$

X=$2z^3-z^2$

Diff.w.r.t.z

$$\frac{dx}{dz} = \frac{d}{dz}(2z^3-z^2)$$

$$\frac{dx}{dz} = 6z^2-2z \quad \text{------(ii)}$$

From (i) and (ii)

$$\frac{dy}{dx} = \frac{dy}{dz} \times \frac{dz}{dx}$$

$\frac{dy}{dx}$ =25z⁴+10/6z²-2z

Higher order differentiation: if a function is differentiating again and again we get first order, second order and so on.

Q.9 If y=$\frac{4}{7}$ x⁵-10x⁴+7x-20 then find d²x/dy²

Sol. y=$\frac{4}{7}$x⁵-10x⁴+7x-20

$\frac{dy}{dx}$ =$\frac{d}{dx}$ ($\frac{4}{7}$x⁵-10x⁴+7x-20)

    =$\frac{4}{7}$.5. x⁴-10.4x³+7

$\frac{dy}{dx}$ =$\frac{20}{7}$x⁴-40x³+7

Again differentiate w.r.t.x

D²y/dx²= $\frac{20}{7}$ 4 x³-40.3x²

Trigonometric Differentiation:

The following are the important formulas we must know to preceding the trigonometric problem on differentiation.

$\frac{d}{dx}$ (Sinx) = cosx

$\frac{d}{dx}$ (Cosx) = -sinx

$\frac{d}{dx}$ (Tanx) = sec²x

$\frac{d}{dx}$ (Cosecx) = -cosecx cotx

$\frac{d}{dx}$ (Secx) = secx. tanx

$\frac{d}{dx}$ (Cotx) = -cosecx. Cotx

INVERSE TRIGONOMETRIC FUNCTION

$\frac{d}{dx}$ (sin⁻¹x) = $\frac{1}{\sqrt{1+x^2}}$

$\frac{d}{dx}$ (cos⁻¹x) = $\frac{-1}{\sqrt{1+x^2}}$

$\frac{d}{dx}$ (cosec⁻¹x) = $\frac{-1}{x\sqrt{x^2-1}}$

$\frac{d}{dx}$ (sec⁻¹x) = $\frac{1}{x\sqrt{x^2-1}}$

$\frac{d}{dx}$ (tan⁻¹x) = $\frac{1}{1+x^2}$

$\frac{d}{dx}$ (cot⁻¹x) = $\frac{-1}{1+x^2}$

Q.10. Differentiate 4tanx+ sec²x w.r.t.x

Sol. Let y=4tanx+ sec²x

    Diff.w.r.t.x

$\frac{dy}{dx} = \frac{d}{dx}$(4tanx+ sec²x)

    $= \frac{d}{dx}$(4tanx) $+ \frac{d}{dx}$(sec²x)

    =4 sec²x+2secx secx. tanx

    =4 sec²x+2sec²x tanx

    = 2sec²x(2+tanx)

11. Diff. $10\sin x + \frac{8}{15}\cos^2 x - 7x^3$

Sol. Let $y = 10\sin x + \frac{8}{15}\cos^2 x - 7x^3$

$= 10\cos x + \frac{8}{15} \cdot 2\cos x\,(-\sin x) - 7.3x^2$

$= 10\cos x - \frac{16}{15}\cos x \sin x - 21x^2$

$= 10\cos x - \frac{8}{15}\sin 2x - 21x^2$ ($\sin 2x = 2\sin x \cos x$)

PROBLEMS FOR PRACTICE

Differentiate the following w.r.t.x

1. $5x^4 + x^3 + \log x + \frac{1}{x} + \frac{1}{x2}$

   (Ans: $4x4 + 3x^2 + \frac{1}{x} - \frac{1}{x2} - \frac{2}{x3}$)

2. $(2x+3)(x+1)$

   (Ans: $4x+5$)

3. $\sqrt{x^2 + 1}$

   (Ans: $\frac{x}{\sqrt{x^2+1}}$)

4. $\frac{2+x^2}{3+x^3}$

   (Ans : $\frac{8x⬚ + 6x^2 + 6x}{(3+x^3)^2}$)

5. $(2 - \frac{2}{x})^3$

   (Ans: $3x^2 - 6 - \frac{12}{x^2} + \frac{24}{x4}$)

6. $\dfrac{x^2+2x+3}{\sqrt{x}}$

(Ans$\dfrac{3}{2}\sqrt{x} + \dfrac{1}{\sqrt{x}} - \dfrac{3}{2x3/2}$)

7. $(3x^2-2)^2(x+1)^3 \ (3(x+1)^2(3x^2-2)(7x^2+4x-2)$

8. $e^{2x}\log(x+1)$

(Ans: e²ˣ+2(x+1)log(x+1)/x+1)

9. $\dfrac{2logx}{x}$

(Ans: $\dfrac{2(1-logx)}{x^2}$

10. $\dfrac{ax+b}{cx+d}$

(Ans: $(\dfrac{ad-bc}{(cx+d)^2}$)

UNIT: VIII

INDEFINITE INTEGRATION

Meaning and Definition

A function f(x) is called a primitive or an integral or antidervative of a function g(x), if g'(x) = f(x).we can find initial value of the function by using the integration. In many business applications we can be use Integration   to find some useful results.

So, if f(x) be a function, then the family of all its primitives is called integral of f(x) and denoted by $\int f(x)dx$

Important formulas

$\int 0dx = c$

$\int K dx = kx + c$

$\int x^n dx = \frac{x^n}{n+1} + c$

$\int \frac{1}{x} dx = logx + c$

$\int e^x \, dx = e^x + c$

$\int a^x \, dx = a^x / loga + c$

$\int (ax + b)^n dx = \frac{(ax+b)^{n+1}}{n+1} + c$

Q.I $\int 4x \, dx$

Sol. $\int 4x \, dx$

$\qquad = 4\int x \, dx$

$\qquad = 4\dfrac{x^2}{2} + C$

Q.2. $\int x(1+x)(1-x) \, dx$

Sol. Let $\quad I = \int x(1+x)(1-x) dx$

$\qquad = \int x(1-x^2) \, dx$

$\qquad = \int (x - x^3) \, dx$

$\qquad = \dfrac{x^2}{2} - \dfrac{x^4}{4} + C$

Q.3 $\int (2x+5)^6 dx$

Sol. Let $\quad I = \int (2x+5)^6 dx$

$\qquad = \dfrac{(2X+5)^{6+1}}{6+1} . 2$

$\qquad = \dfrac{1}{7}(2X+5)^7 . 2$

$\qquad = \dfrac{2}{7}(2X+5)^7 + C$

Q.4 $\int (x - \dfrac{1}{x})^3 dx$

Sol. Let $I = \int (x - \dfrac{1}{x})^3 dx$

$\qquad = \int (x^3 - 3x + \dfrac{3}{x} - \dfrac{1}{x^3}) \, dx$

$$=\int x^3 dx - 3\int x\,dx + 3\int \frac{1}{x}dx - \int \frac{1}{x^2}dx$$

$$=\frac{x^4}{4} - 3\frac{x^2}{2} + 3\log x + \frac{1}{3x^3} + C$$

Some standard results used in trigonometric integration

$\int sinx\ dx = -\cos x + c$

$\int cosx\ dx = \sin x + c$

$\int sec^2 x dx = \tan x + c$

$\int cosec^2 x dx = -\cot x + c$

$\int secx.\ tanx\ dx = \sec x + c$

$\int cosecx.\ cotx\ dx = -\mathrm{cosec} x$

Q.5 $\int (sinx + x^3)dx$

Sol. *Let I* $= \int (sinx + x^3)dx$

$$=\int sinx\ dx + \int x^2 dx$$

$$= -\cos x + \frac{x^3}{3} + c$$

Q.6 $\int (cos2x + cosec^2 x + sec^2 x + x^3)\ dx$

Sol. Let I $=\int (cos2x + cosec^2 x + sec^2 x + x^3)\ dx$

$=\int cos2xdx + \int cosec^2 xdx + \int sec^2 xdx + \int x^3 dx$

$= \frac{sin2x}{2} - \cot x + \tan x + \frac{x^4}{4} + c$

Methods of integration

1. Integration by substitute

2. Integration by partial fraction

3. Integration by parts.

Integration by substitute method

Under this method some difficult integration can be can be easily processed by help of using third variable as the supportive variable. The following illustration are explained the situation of substitution method.

Q.7 $\int (4x + 5)^6.dx$

Sol. Let I $= \int (4x + 5)^6 dx$

Put 4x+5 = t

4dx = dt

Now, I $= \frac{1}{4} \int t^6 dt$

$= \frac{1}{4} \frac{t^7}{7} + c$

$= \frac{1}{28} t^7 + c$

$= \frac{1}{28}(4x+5)^7 + c$

Q.8. Integrate $\int \frac{x^3}{(x^2+1)^3} dx$

Sol. Let I $= \int \frac{x^3}{(x^2+1)^3} dx$

Put $x^2+1 = t$

$X^2 = t-1$

$2xdx = dt$

$I = \int \frac{x^2.x}{(x^2+1)^3}dx$

$I = \int \frac{(t-1)dt}{t^3}$

$= \int \frac{1}{t^2}-\frac{1}{t^3}dt$

$= -\frac{1}{t}+\frac{1}{2t^2}+c$

$= -\frac{1}{x^2+1}+\frac{1}{2(x^2+1)^2}+c$

Q.9 Evaluate$\int x^3 \sin x^4 dx$

Sol. Let $I = \int x^3 \sin x^4 dx$

Put $x^4 = t \Rightarrow 4x^3dx = dt$

$I = \frac{1}{4}\int \sin t\, dt$

$= \frac{1}{4}(-\cos t)+c$

$= -\frac{1}{4}\cos t +c$

Integration by Parts

Under this method used when two or more function are in product form we can apply this method to solve the problems easily. The different function may be trigonometry, logarathimic, exponential, Inverse and algebraic. It is arranged in the sequence of ILATE (Inverse, log, Algebraic, trigonometry, Exponential)

$\int uv dx = u\int v dx - \int(\frac{d}{dx}(u).\int v dx)\,dx$, the functions U and V should we arranged the order of **ILATE** rule

ILATE-rule means when we have different function, before start integration by parts we should have to arrange i.e. I-inverse, L-lograthimics, A-Algebric, T-Tirigonometric, E-exponential functions.

Q.10 Integrate $\int x^2 sinx dx$

Sol. Let I $=\int x^2 sinx dx$

$= x^2\int sinx\,dx - \int(\frac{d}{dx}(x^2).\int sinx dx)dx$

$=- x^2cosx + \int 2x\,cosx dx$

$=- x^2cosx + 2(x\,sinx - \int 1.sinx dx)$

$=- x^2cosx + 2(x\,sinx - cosx) + c$

Q.11 Integrate $\int e^x x^2 dx$

Sol. Let I $= \int e^x x^2 dx$

$=\int x^2 e^x dx$

$= x^2\int e^x - \int(\frac{d}{dx}(x^2)\int e^x dx)dx$

$= x^2 e^x - \int 2x e^x dx$

$= x^2 e^x - 2\int x e^x dx$

$= x^2 e^x - 2(x e^x - \int e^x dx)$

$= x^2 e^x - 2(x e^x - e^x) + c$

Q.12 Integrate $\int xlogx\,dx$

Sol. Let I $= \int xlogx\,dx$

$\qquad = \int logx.x.dx$

$\qquad = logx\int x\,dx - \int(\frac{d}{dx}(logx)\int x\,dx)dx$

$\qquad = logx.\frac{x^2}{2} - \frac{1}{2}\int x\,dx$

$\qquad = logx\frac{x^2}{2} - \frac{1}{2}\frac{x^2}{2} + c$

Q.13. Integrate $\int e^x sinx\,dx$

Sol. Let I $= \int e^x sinx\,dx$

$\qquad = e^x\int sinx\,dx - \int\frac{d}{(dx)}(e^x)\int sinx\,dx$

$\qquad = -e^x cosx + \int e^x cosx\,dx$

$\qquad = -e^x cosx + e^x sinx - \int e^x sinx\,dx$

I $= -e^x cosx + e^x sinx - I$

2I $= -e^x cosx + e^x sinx$

I $= \frac{1}{2}(-e^x cosx + e^x sinx)$

Integration by Partial fraction

If the fractional integral part with linear factors in denominator then we apply this method to get the solution easily. The following examples helps to understand this method.

Q.13 Integrate $\int \frac{1}{(x+2)(x-4)} dx$

Sol. Let $I = \int \frac{1}{(x+2)(x-4)} dx$

$$\frac{1}{(x+2)(x-4)} = \frac{A}{(X+2)} + \frac{B}{(X-4)}$$

Comparing with numerator

$1 = A(x-4) + B(x+2)$

Put $x=4$, $1 = 6B \Rightarrow B = \frac{1}{6}$

Put $x=-2$, $1 = -6A \Rightarrow A = \frac{-1}{6}$

Now, $I = \frac{-1}{6} \int \frac{dx}{x+2} + \frac{1}{6} \int \frac{dx}{x-4}$

$I = \frac{-1}{6} \log(x+2) + \frac{1}{6}\log(x-4) + c$

Q.14. Integrate$\int \frac{(2x-1)dx}{(x-3)(x+1)}$

Sol. $I = \int \frac{(2x-1)dx}{(x-3)(x+1)}$

Let $\frac{(2x-1)}{(x-3)(x+1)} = \frac{A}{X-3} + \frac{B}{X+1}$

Comparing with numerator

$2x-1 = A(x+1) + B(x-3)$

Put $x=-1$, $-3 = -4B \Rightarrow B = \frac{3}{4}$

Put $x= 3$, $5 = 4A \Rightarrow A = \frac{5}{4}$

Now, $I = \frac{5}{4}\int \frac{dx}{x-3} + \frac{3}{4}\int \frac{dx}{x+1}$

$\qquad = \frac{5}{4}\log(x-3) + \frac{3}{4}\log(x+1) + c$

PROBLEMS FOR PRACTICE

Q.1 $\int (5e^x + \frac{3}{x^2})\,dx$

Ans: $5e^x - \frac{3}{x}$

Q.2 $\int \frac{1}{2}\sec^2 x\,dx$

Ans: $\frac{1}{2}\tan x + c$

Q.3 $\int \frac{(1+X)^2}{\sqrt{X}}\,dx$

Ans: $2\sqrt{x} + \frac{4}{3}x^{3/2} + \frac{2}{5}x^{5/2} + c$

Q.4 $\int (\tan x + \cot x)^2\,dx$

Ans: $\tan x - \cot x + c$

Q.5 $\int \frac{x^3}{(X+1)^2}\,dx$

Ans: $\frac{x^2}{2} - 2x + 3\log(x+1) + \frac{1}{x+1} + c$ (Hint: put $x+1=t$)

Q.6 $\int x.\log x\,dx$

Ans: $\frac{x^2}{2}\log x - \frac{1}{4}x^2 + c$ (Hint: Use by parts method)

Q.7 $\int x^3 e^d x$

Ans: $(x^3 - 3x^2 + 6x - 6)e^x + c$

Q.8 $\int x \sin 2x \, dx$

Ans: $-\frac{x}{2}\cos 2x + \frac{1}{4}\sin 2x + c$

Q.9 $\int \frac{X-1}{(X+1)(X-2)} dx$

Ans: $\frac{2}{3}\log(x+1)\frac{1}{3}\log(x-2)+c$

Q.10 $\int \frac{2x+1}{(x+1)(x-2)} dx$

Ans: $\frac{1}{3}\log(x+1)+\frac{5}{3}\log(x-2)+c$

BASIC AIGEBRIC FORMULAS (Thinks to remember)

$(a+b)^2 = a^2 + 2ab + b^2$

$(a-b)^2 = a^2 - 2ab + b^2$

$a^2 - b^2 = (a+b)(a-b)$

$a^3 + b^3 = (a+b)(a^2 - ab + b^2)$

$a^3 - b^3 = (a-b)(a^2 + ab + b^2)$

$(a+b)^3 = a^3 + 3a^3b + 3ab^2 + b^3$ (or) $a^3 + b^3 + 3ab(a+b)$

$(a-b)^3 = a^3 - 3a^2b + 3ab^2 - b^3$ (or) $a^3 - b^3 - 3ab(a-b)$

$(a+b+c)^2 = a^2 + b^2 + c^2 + 2ab + 2bc + 2ca$

$(x+a)(x+b) = x^2 + (a+b)x + ab$

$a^3 + b^3 + c^3 - 3abc = (a+b+c)(a^2 + b^2 + c^2 - ab - bc - ca)$

$a^3 + b^3 + c^3 = 3abc$ (if $a+b+c=0$)

LAWS OF EXPONENT

$a^m \times a^n = (a)^{m+n}$

$a^m \div a^n = (a)^{m-n}$

$(a^m)^n = (a)^{mn}$

$a^{-1} = \dfrac{1}{a}$

$\left(\dfrac{a}{b}\right)^0 = 1$

$a^m \times b^m = (ab)^m$

Types of Ratios

1. Duplicate ratio of a: b $= a^2 : b^2$

2. Sub duplicate ratio of a: b $= \sqrt{a} : \sqrt{b}$

3. Triplicate ratio of a: b $= a^3 : b^3$

4. Sub triplicate ratio of a:b $= \sqrt[3]{a} : \sqrt[3]{b}$

PROFIT AND LOSS

Profit of Gain = sp – cp

Loss = cp-sp

Sp (on gain) = cp $\left(\dfrac{100+p\%}{100}\right)$

Sp (on loss) = cp $\left(\dfrac{100-l\%}{100}\right)$

P% $= \dfrac{p}{cp}$ X 100

$L\% = \frac{L}{CP} X 100$

Discount is allowed on basic price (or) market price (or) list price.

Sp (on discount) = mp $(\frac{100-d\%}{100})$

D% = $\frac{Discount}{mp} X 100$

SIMPLE INTEREST AND COMPUND INTEREST

$SI = \frac{PXRXT}{100}$

P: Principal

R: Rate of interest

T: time period

S.I: Simple Interest

$P = \frac{SI \ X \ 100}{RT}$; $T = \frac{SI \ X \ 100}{PR}$; $R = \frac{SI \ X \ 100}{PT}$

Amount on compound (A) = P $(1 + \frac{R}{100})^n$

Compound interest = A-P

$$= P (1 + \frac{r}{100}) - p$$

$$C.I = P [(1 + \frac{r}{100}) - 1]$$

Where, R=Rate of interest, n=Time period, P-Principal amount

PROPERTIES OF LOGRITHIMS

(I) log (mn) =log m + log n

(ii) Log m^n = n log m

(iii) Log $\frac{m}{n}$ = log m - log n

MATRICES AND DETERMINANT

Properties of determinants (P)

P1- Let A=$[a_{ij}]$ be a square matrix of order n, then the sum if the product of elements of any row or column with their cofactors is always equal.

P2- Let A=$[a_{ij}]$ be a square matrix of order n, then the sum of the product of elements of any row(column) with corresponding elements of some other column(row) is zeros.

P3-Let A=$[a_{ij}]$ be a square matrix of order n, and $|A| = \Delta$, The value of determinant remains unchanged if rows and columns are interchanged.

P4- Let A=$[a_{ij}]$ be a square matrix of order n, and $|A| = \Delta$,If any two rows or columns of a determinants are interchanged, then value of determinant is change in sign.

P5- Let A=$[a_{ij}]$ be a square matrix of order n, and $|A| = \Delta$,If any two rows or columns of a determinants are identical, then its value is zero.

P6- Let A=$[a_{ij}]$ be a square matrix of order n, and $|A| = \Delta$, If each element of a row or column of a determinant is multiplied by a constant k, the value of determinant is also get multiplied by k.

P7- Let $A = [a_{ij}]$ be a square matrix of order n, and $|A| = \Delta$, If each element of a row or column if a determinant is expressed as a sum of two or more terms, then the determinants can be expressed as the sum of two or more determinants.

P8- Let $A = [a_{ij}]$ be a square matrix of order n, and $|A| = \Delta$, If each elements of a row or columns of a determinant is multiplied by same constant and then added to any row or column, then value of determinant remains same.

Inverse of matrix A is given by, $A^{-1} = \frac{1}{|A|} \text{adj} A$

Cramer's Rule

Consider the following equations: $a_1x + b_1y = c_1$ and $a_2x + b_2y = c_2$

$$\Delta = \begin{vmatrix} a_1 & b_1 \\ a_2 & b_2 \end{vmatrix} \quad \Delta x = \begin{vmatrix} a_1 & b_1 \\ a_2 & b_2 \end{vmatrix} \text{ and } \Delta y = \begin{vmatrix} a_1 & b_1 \\ a_2 & b_2 \end{vmatrix}$$

We can classify the solution of the given equations in three different solution on the basis of value of Δ Δx and Δy

i. If $\Delta \neq 0$, then the system of equation has unique solution.

ii. If $\Delta = 0$, and, $\Delta x \neq 0$, $\Delta y \neq 0$, then the equation has No solution system.

iii. If $\Delta = 0$, $\Delta x = 0$ and $\Delta y = 0$, then the equation has infinite number of solutions.

Indefinite Integration | 195

Similarly we can apply in three variable equations

$a_1x+b_1y+c_1z=d_1$, $a_2x+b_2y+c_2z=d_2$ and $a_3x+b_3y+c_3z=d_3$

Here, $\Delta = \begin{vmatrix} a_1 & b_1 & c_1 \\ a_2 & b_2 & c_2 \\ a_3 & b_3 & c_3 \end{vmatrix}$, $\Delta x = \begin{vmatrix} d_1 & b_1 & c_1 \\ d_2 & b_2 & c_2 \\ d_3 & b_3 & c_3 \end{vmatrix}$,

$\Delta y = \begin{vmatrix} a_1 & d_1 & c_1 \\ a_2 & d_2 & c_2 \\ a_3 & d_3 & c_3 \end{vmatrix}$ and, $\Delta z = \begin{vmatrix} a_1 & b_1 & d_1 \\ a_2 & b_2 & d_2 \\ a_3 & b_3 & d_3 \end{vmatrix}$

If the equation has unique solution system then we can find the x,y and z by the following equations

$X = \frac{\Delta x}{\Delta}$, $y = = \frac{\Delta y}{\Delta}$, $z = \frac{\Delta z}{\Delta}$

Matrix method for the solution of a Homogeneous system of simultaneous linear equations

To obtain the system of equations consider the following equations

$a_1x+b_1y+c_1z=d_1$, $a_2x+b_2y+c_2z=d_2$ and $a_3x+b_3y+c_3z=d_3$

We can rewrite in matrix method as follows

$\begin{bmatrix} a_1 & b_1 & c_1 \\ a_2 & b_2 & c_2 \\ a_3 & b_3 & c_3 \end{bmatrix} \begin{matrix} x \\ [y] = \\ z \end{matrix} \begin{bmatrix} d_1 \\ d_2 \\ d_3 \end{bmatrix}$

A   X =   B

(i) If $|A| \neq 0$, then the system of equations is consistent with unique solution and we can find the value of X by using the identity $X = A^{-1}B$.

(ii) If $|A| = 0$, then the given system of equations is either inconsistent or it has infinitely many solution. It can be confirmed by the following workings

a) If (adjA) B$\neq$0, then the system will be inconsistent or no solution system.

b) If (adjA) B=0, then the system will be consistent with infinitely many solutions.

Limits: Proerties and furmula

- $\lim_{x \to a}(f(x) \pm g(x)) = \lim_{x \to a}f(x) \pm \lim_{x \to a}g(x)$

- $\lim_{x \to a}(f(x) \cdot g(x)) = \lim_{x \to a}f(x) \cdot \lim_{x \to a}g(x)$

- $\lim_{x \to a}(f(x)/g(x)) = \lim_{x \to a}f(x)/\lim_{x \to a}g(x)$

- $\lim_{x \to a}\dfrac{x^n - a^n}{x - a} = na^{n-1}$

- $\lim_{x \to 0}\dfrac{e^x - 1}{x} = 1$

- $\lim_{x \to 0}\dfrac{1}{x} = \infty$

Differentiation: Properties and formula

- $\dfrac{d}{dx}(c) = 0$

- $\dfrac{d}{dx}(x^n) = nx^{n-1}$

- $\frac{d}{dx}(u^n) = nu^{n-1}\frac{d(U)}{dx}$

- $\frac{d}{dx}(U \pm V) = \frac{d}{dx}(U) \pm \frac{d}{dx}(V)$

- $\frac{d}{dx}(UV) = U\frac{d}{dx}(v) + V\frac{d}{dx}(U)$

- $\frac{d}{dx}(\frac{U}{V}) = \frac{V\frac{dU}{dx} - u\frac{dv}{dx}}{V^2}$

- $\frac{d\,(logx)}{dx} = \frac{1}{x}$

- $\frac{d(e)^x}{dx} = e^x$

- Chain rule: $\frac{dy}{dx} = \frac{dx}{dt} \times \frac{dt}{dy}$

- Successive differentiation of a function Y is known as first order, second order, third order Denoted as $\frac{dy}{dx}$, $d^2y/dx^2, d^3y/dx^3$ ...... respectively.

- $\frac{d}{dx}(Sinx) = cosx$

- $\frac{d}{dx}(Cosx) = -sinx$

- $\frac{d}{dx}(Tanx) = sec^2x$

- $\frac{d}{dx}(Cosecx) = -cosecx\ cotx$

- $\frac{d}{dx}(Secx) = secx.\ tanx$

- $\frac{d}{dx}(Cotx) = -cosecx.\ Cotx$

INVERSE TRIGONOMETRIC FUNCTION

- $\frac{d}{dx}(\sin^{-1}x) = \frac{1}{\sqrt{1+x^2}}$

- $\frac{d}{dx}(\cos^{-1}x) = \frac{-1}{\sqrt{1+x^2}}$

- $\frac{d}{dx}(\csc^{-1}x) = \frac{-1}{x\sqrt{x^2-1}}$

- $\frac{d}{dx}(\sec^{-1}x) = \frac{1}{x\sqrt{x^2-1}}$

- $\frac{d}{dx}(\tan^{-1}x) = \frac{1}{1+x^2}$

- $\frac{d}{dx}(\cot^{-1}x) = \frac{-1}{1+x^2}$

Integration Formulas:

- $\int 0\,dx = c$

- $\int K\,dx = kx + c$

- $\int x^n\,dx = \frac{x^n}{n+1} + c$

- $\int \frac{1}{x}\,dx = \log x + c$

- $\int e^x\,dx = e^x + c$

- $\int a^x\,dx = a^x / \log a + c$

- $\int (ax+b)^n\,dx = \frac{(ax+b)^{n+1}}{n+1} + c \int \sin x\,dx = -\cos x + c$

- $\int \cos x\,dx = \sin x + c$

- $\int \sec^2 x\,dx = \tan x + c$

- $\int \csc^2 x\,dx = -\cot x + c$

- $\int \sec x.\tan x\,dx = \sec x + c$

- $\int \csc x.\cot x\,dx = -\csc x + c$

PONDICHERRY UNIVERSITY
B.COM. DEGREE EXAMINATION APRIL 2013
BUSINESS MATHEMATICS

Time: Three hours   maximum: 100 marks

SECTION A- (10X3=30)

1. What is meant by ratios?

2. Define the term percentage

3. Give three differences between trade discount and cash discount.

4. Write short notes on the following:

    a. Selling price

    b. Cost price

5. What is meant by compound interest?

6. Explain the concept of shares

7. Define matrix

8. Find the inverse of $\begin{bmatrix} 2 & 2 \\ 3 & 5 \end{bmatrix}$, if exists

9. Explain the meaning of business variable.

10. What is a monotone function variable?

SECTION B-(5X6=30)

11. Two numbers are in the ratios of 3:4, if 6 are added to each of the term, then the new ratio become 4:5. Find the numbers.

12. Find the mean proportion of 9 and 4, the third proportion to 4 and 6.

13. Write a note commission and brokerage.

14. Find the sum which will yield an interest of ₹324 in 3 years at 4% P.A simple interest.

15. Find the compound interest on ₹ 7200 for 2 years at $4\frac{1}{2}$ %p.a.

16. Discuss the various types of matrices.

17. Compute determinant $A = \begin{bmatrix} 2 & 3 & -4 \\ 0 & -4 & 2 \\ 1 & -1 & 5 \end{bmatrix}$

18. Explain the function of inverse and absolute value.

SECTION C-(2X20=40)

19. What are the various methods to calculate simple interest and compound interest?

20. Given two Matrices A and B where

$A = \begin{bmatrix} 1 & -1 & 0 \\ 2 & 3 & -4 \\ 0 & 1 & 2 \end{bmatrix} B = \begin{bmatrix} 2 & 2 & -4 \\ -4 & 2 & -4 \\ 2 & -1 & 5 \end{bmatrix}$

Verify that AB=BA=6I. Using this result solve the set of linear equations
X-y =1, 2x+3y-4z=2 and y+2z=5

The value of diamond varies as into 5 square of its weight. A diamond varies as the squares of its weight. A diamond is broken into 5 pieces, the weights of which are in the ratio 1:2:3:4:5. If the resulting loss is ₹ 85,000 find the value of original diamond. Also calculate the value of a diamond whose weight that of the original diamond.

PONDICHERRY UNIVERSITY
B.COM. DEGREE EXAMINATION
APRIL2014
BUSINESS MATHEMATICS

Time: Three hours   maximum: 100 marks

SECTION A- (10×3=30)

1. Define ratio.
2. Find the mean proportion of 9 and 4 and the third proportion between 4 and 6
3. Write a short note on trade discount.
4. What you understand by brokerage?
5. Calculate simple interest on ₹9000 at 5% per annum for 4 years.
6. What do you understand by bonus shares?
7. Define matrix
8. If A=$\begin{bmatrix} 1 & 2 \\ -3 & 4 \end{bmatrix}$ and B=$\begin{bmatrix} 5 & -6 \\ 7 & 8 \end{bmatrix}$, then find A+ B
9. Define function.
10. Write a short note on cost function.

SECTION B-(5X6=30 Marks)

Answer any FIVE questions.

1. The monthly salaries of two persons are in the ratio of 3:5. If each receives an increase of ₹ 20 in monthly salary the ratio is altered to 13:21. Find their salaries.
2. Distinguish between trade discount and cash discount.

3. Find the compound interest on the sum of ₹6280 for one year and seven months at the rate of 8% per annum reckoned yearly.

4. Find the inverse of $A = \begin{bmatrix} -6 & -12 \\ -8 & -8 \end{bmatrix}$

5. Define the following terms.
   a) Constants
   b) Variables
   c) Domain.

6. Briefly explain the problems on commission and brokerage.

7. What is meant by percentage? How will you compute it?

8. Briefly explain the types of matrices.

SECTION C - (2X20=40 Marks)

9. Discuss the various types of functions.

10. A person borrows 2500 at 10% simple interest for 2 years. He immediately lends this money out at compound interest at the same rate and for the same time. What is his gain at the end of 2 years?

11. Find the inverse of $A = \begin{bmatrix} 4 & 0 & 2 \\ 2 & 10 & 2 \\ 3 & 9 & 1 \end{bmatrix}$

PONDICHERRY UNIVERSITY
B.COM. DEGREE EXAMINATION
NOVEMBER-2014
BUSINESS MATHEMATICS

Time: Three hours   maximum: 100 marks

SECTION A- (10×3=30)

1. **Define** variation.
2. **What is continued proportion?**
3. Write note on trade discount
4. Differentiate the commission from brokerage.
5. What is EMI?
6. What is meant by dividend?
7. Define matrix.
8. Write short note on linear equations.
9. What is an algebraic function?
10. What do you mean by variables?

SECTION B-(5×6 = 30)

11. The volume of a gas varies directly as the absolute temperature and inversely as the pressure. When the pressure is 15 units and the temperature is 260 units, the volume is 200 units. What will be the volume when the pressure is 18 units and the temperature is 195 units?

12. The monthly incomes of two persons are in the ratio 6:7 and their monthly expenditure is in the ration 11:13, if each saves Rs 50 per month, find their monthly incomes.

13. An agent receives a fixed salary and a commission on orders booked. In two successive months he obtained orders to the value of Rs 35000 and Rs 50000 respectively and received Rs 2800 and Rs 3400 including his commission and salary. Find the rate of commission and his fixed salary.

14. Find the interest on Rs 1000 for 10 years at 4% P.a. the interest being paid quarterly.

15. State the functions of stock exchanges.

16. Find the adjoint of $\begin{bmatrix} 3 & 1 & 2 \\ 2 & 2 & 5 \\ 4 & 1 & 0 \end{bmatrix}$

17. Write short note on:

    a. Constants

    b. Variables

    c. Domain

18. A survey shows that 57% of Indian likes coffee whereas 75% like tea. What do you say about % of Indians who like both coffee and tea?

SECTIONS C (2×20=40)

19. Find the inverse of $\begin{bmatrix} 1 & 0 & -4 \\ -2 & 2 & 5 \\ 3 & -1 & 2 \end{bmatrix}$

20. Two vessels A and B contain mixtures of spirit and water. A mixtures of 3 parts from A and 2 parts from B is found to contain 29% of spirit, and a mixture of 1 part from A and 9 parts from B is found to contain 34% of spirit. Find the % of spirit in A and B.

21. List and explain any ten business variables.

PONDICHERRY UNIVERSITY
B.COM. DEGREE EXAMINATION APRIL/MAY 2016
BUSINESS MATHEMATICS

Time: Three hours   Maximum: 100 Marks

SECTION A (10X3 = 30 marks)

1. What is ratio?
2. Determine joint variation.
3. What is commission?
4. Find which is larger ration? 4:10 and 6:7
5. Fin the simple interest and the amount Rs 5000 at 4 ½% for 1 ½ years.
6. Define Matrix.
7. Define Dividend.
8. What is Polynomial function?
9. What is production function?
10. Define Linear equation

SECTION B - (5X6 =30 marks)

11. Express the following ratios in decimals
    a)2:10   b)100:50
12. Divide Rs 510 between A, B and C so that A gets 2/3 of what B gets and b gets ¼ what C gets. Finds the share of each.
13. Discuss the various types of matrices.
14. Evaluate $\int \frac{4x^7-3x^3-5x^2}{x^4}\, dx$ (As per the new syllabus this question is irrelevant)
15. Evaluate $\begin{bmatrix} 4 & 1 & 3 \\ 2 & 0 & -6 \\ 5 & -7 & 9 \end{bmatrix} \times \begin{bmatrix} 5 & -2 & 0 \\ 1 & 6 & 8 \\ 3 & 4 & 7 \end{bmatrix}$

16. A sum of money invested at compound interest amount to Rs 21632 in 2 years and to Rs 22498 in 3 years. Find the rate of interest and the sum invested.

17. Describe the various types of functions.

18. Find the inverse of A. If A= $A = \begin{bmatrix} 1 & 0 & -1 \\ 3 & 4 & -5 \\ 0 & -6 & -7 \end{bmatrix}$

SECTION C - (2X20= 40 marks)

19. The monthly salaries of two persons are in the ratio of 3:5. If each receives an increase of Rs 200 in monthly salary the ratio is altered to 13:21. Find their salaries.

20. Published price of a book is Rs 15. The publisher offers a trade discount of 16% and one book free for every 20 books purchased. Besides a cash discount of 4% is offered. How much does the purchaser get for each book?

21. Solve the following equation by Crammer's rule.
    $3x+2y = 8$ and $5x-3y = 7$

22. Solve the following simultaneous equation by matrix method.
    $x_1+2x_2+x_3 = 2$
    $2x_1+2x_3+x_4 = 6$
    $4x_2+3x_3+2x_4 = -1$
    $-x_1+6x_2-x_3-x_4 = 2$

Time: Three hours   Maximum: 100 Marks

SECTION A -(10X3 = 30 marks)

1. Define Ratio.
2. Express the following rations in decimals:
   a) 2:100   b) 1000:50   c) 3:12
3. Find the value of x in the proportion
   $(x+5): (3x-6) = 3:8$
4. Find simple interest on
   a) Rs 20000 for one year, three months and 15 days at 24% p.a.
   b) Rs 10000 for 73 days at 20% p.a.
   c) Rs 2500 for 15 months at 15% p.a.
5. At what time will be Rs 7500 amount to Rs 8670 at 6% p.a. compound Interest?
6. Write short notes on:
   a) Trade discount
   b) Cash discount
   c) Bonus shares
7. Explain the following terms:
   a) Equity shares
   b) Dividend
   c) Define function
8. Define Matrices.
9. If $A = \begin{bmatrix} 4 & -1 \\ -7 & 2 \end{bmatrix}$ $B = \begin{bmatrix} 2 & 1 \\ 7 & 4 \end{bmatrix}$ show that $AB = BA$

10. Briefly explain the explicit and algebraic functions.

SECTION B (5X 6 =30 marks)

11. Find the adjoint of the Matrix $A = \begin{bmatrix} 4 & -1 \\ 3 & 2 \end{bmatrix}$

12. Briefly explain the following terms.
    a) Profit function
    b) Production function
    c) Utility function
    d) Consumption function.

13. Discuss the various types of matrix.

14. If a+b: a-b = 7:3, find the value of a: b.

15. At what rate percent per annum the compound interest on Rs 9000 amounts to Rs amounts to 1327.70 in 4 years?

16. The Price of an article in 1983 was 22 ½ % higher than in 1982. In 1984 the price was 35% higher than 1983. What percentage advance was the 1984 price on the 1982 price?

17. The ratio of the prices of 2 bicycles was 16:23. Five years later when the price of the first had increased by 20% and that of the second by Rs 617, the ratio of their prices became 13:22. What was the original price of the bicycles?

18. Write short notes on
    a) Commission and brokerage.
    b) Supply and demand function
    c) Rules of proportion.

SECTION C (2X20 = 40 marks)

19. A man gave 3/8 of his property to one son and 30% of the remainder to another. He then distributed the remaining property among three charities in the proportion 2:5:7. The difference of his son's share was Rs 4200. What was the value of his property and how much did each charity receive?

20. If $A = \begin{bmatrix} 1 & -3 \\ -2 & 4 \end{bmatrix}$ $B = \begin{bmatrix} 3 & 0 & 1 \\ 6 & -2 & -7 \end{bmatrix}$ and $C = \begin{bmatrix} 1 & -3 \\ 6 & 5 \\ 0 & -2 \end{bmatrix}$ verify that $A(BC) = (AB)C$

21. Write notes on:
    a) Total revenue function
    b) Rational and irrational function
    c) Define constants, variables
    d) Continued, Direct and Inverse proportion.
    e) Define cost price and selling price.

Time: Three hours   Maximum: 100 Marks

SECTION A - (10X3 = 30 marks)

1. Define Ratio
2. Find the sum if 15% of the sum is Rs 343.50.
3. The cash price of a machine is Rs 15200. The trade discount and cash discount are 5% and 20% respectively. Find the catalogue price.
4. A bank paid Rs 3454 for a bill of Rs 3500 drawn on 1st April at 5 months date. On what day the bill as discounted if the rate of interest is 8% p.a.
5. Find the period in which Rs 1500 will yield Rs 270 as simple interest at 12% p.a.
6. What are bonus shares?
7. Define square matrix.
8. $A = \begin{bmatrix} 1 & 2 \\ 3 & 4 \end{bmatrix}$, $B = \begin{bmatrix} 1 & 0 \\ 2 & -3 \end{bmatrix}$, $C = \begin{bmatrix} 1 & -1 \\ 0 & 1 \end{bmatrix}$

   Then show that $A(B+C) = AB + AC$
9. What are constants?
10. Write a note on supply and demand functions.

SECTION B (5 x 6 =30 marks)

11. The price of an article in 2013 was 22 ½ % higher than in 2012. In 2014 the price was 35% higher than in 2013. What percentage advance was the 2014 price on the 2012 price?

12. How much tea at Rs 9 per kg must be mixed with 100 kg of superior tea at Rs 13.50 per kg to give an average price of Rs 11 per kg?

13. The bankers gain on a bill due after three months at 8% p.a. is Rs 25. Find the true discount and the face value of the bill

14. The listed price of an article is Rs 125. The seller offers a trade discount of 10% and one article free for every 10 articles. In addition, a cash discount of 5% also offered. How much does the seller get for each article?

15. At what time will Rs 1000 amount to Rs 2000 at 8% p.a. compound interest?

16. Find the principal for which the difference between simple interest and compound interest for 3 years at Rs 80 at 8% p.a.

17. The demand function faced by a firm is p=500-0.2x and is cost function is C=25x+10000 (p= Price= output and c= cost). Find the output at which the profit of the firm are maximum. Also find the price to will charge.

18. Find the equilibrium price and quantity for the demand and supply functions
$Q_d = 4- 0.05p$ and $Q_s =0.8+.11p$

SECTION C -(2X20 = 40 marks)

19. Raja, Ravi and Rammu form a business with a capital of Rs 63000. Of this Raja contributed Rs 26400. Ravi contributed Rs 22200 and Rammu the balance. After four months, Raja withdrew Rs 5400 and Ravi and Rammu introduced Rs 2700

each. At the end of the year, the profit to be divided was 15% of the total capital.

20. Explain the different types of functions.

21. There are 2 families, A and B. In family A, 2 men, 3 women and a child are there. In family B, a man, a woman and two children are there. The recommended daily diet for calorie is man 2400, woman 1900 and a child 1800 and for proteins, man-55 gm, woman-45gm and child-33gm. Represent the above by matrices. Calculate total weekly requirement of calories and proteins for both families.

Solution to Business Mathematics
Pondicherry University
B.Com Degree Examination April– 2015

SECTION A

1. Ratio is a simplest fraction derived from the comparison of similar character or nature of any quantitative measurable things. So, it converts the complicities terms into simpler term to compare or study for any necessities.

2. No. of students joined in NCC $=30$
   No. of students joined in NSS $=20$

   Total no. of students in the class $=60$

   So, %of students not joined either NCC or NSS

   $=\frac{60-(30+20)}{60}$ X100

   $=\frac{10}{60}$X100

   $=16.67$ %(Approx)

3. Cash discount means the discount offered by a seller to a consumer or customer on cash purchase on a special occasion or seasonable occasion.

4. Brokerage is an income charged by broker when he completes a financial dealing between two parties. The financial dealing may be relates to making link between buyer and seller regarding Share, Debentures, Land or Assets etc.

5. Sum invested = Rs 6000

   Rate of interest = 10%

   No. of years = 3years

   Simple interest $=\dfrac{PXRXT}{100}=\dfrac{6000X10X3}{100}=100$

6. Share is part of capital of a company. When company needs a large capital to start its business it decides to raise the fund through the issue of share. Each share has a nominal value which accumulates the total capital required to the company.

7. A square matrix which has real value as unit or one is called unit matrix. The unit matrix can be in any order but the value should have as one. Sometimes it also called as Identity Matrix.

   For example, $\begin{bmatrix} 1 & 0 \\ 0 & 1 \end{bmatrix}$ and $\begin{bmatrix} 1 & 0 & 0 \\ 0 & 1 & 0 \\ 0 & 0 & 1 \end{bmatrix}$ are the unit matrices of 2 order, 3 order respectively.

8. $A = \begin{bmatrix} 2 & 3 \\ 10 & 15 \end{bmatrix}$ $B = \begin{bmatrix} 4 & 7 \\ 8 & 24 \end{bmatrix}$

   $$A+B = \begin{bmatrix} 2 & 3 \\ 10 & 15 \end{bmatrix} + \begin{bmatrix} 4 & 7 \\ 8 & 24 \end{bmatrix}$$

   $$= \begin{bmatrix} 2+4 & 3+7 \\ 10+8 & 15+24 \end{bmatrix}$$

   $$= \begin{bmatrix} 6 & 10 \\ 25 & 32 \end{bmatrix}$$

9. Variables are unknown value which varied accordingly the business environment. In business sales, production, production cost,

expenses etc.are variables. It generally denotes x, y or z.

10. Supply function is the studies the functional relationship between physical inputs and physical output of a commodity $Q_x=f(L,K)$ L (Labour),K (Capital).

Demand function shows the relationship between demand for a commodity and its various determinants. It shows how demand for a commodity is related to say own price of the commodity or income of the consumer or other determinants.

SECTION B

11. Ratio of incomes of brothers $= 9:7$

Ratio of their amount spends $=6:5$

So, we can assumes the income as 9x and 7x

And the expenditure can assumes as 6y and 5y

We can represent the information in the following equations

9x-6y =1500 ------- (i)

7x-5y =1000 -------- (ii)

Eqn (i)x7   63x-42y =10500 ----(iii)

Eqn(ii)x9   63x -45y=9000   ----(iv)

Eqn (iii) - (iv)   3y = 1500

Y = 500

Putting y=500 in equation (i) x=500

Incomes of two brothers are Rs4500 and Rs3500.

12. Cash Discount:

Cash discount means the discount offered by a seller to a consumer or customer on cash purchase on a special occasion or seasonable occasion. In order to increase the sales the seller offers the discount. It is not in regular nature. Rate of cash discount will changed on the nature of goods and value of goods.

Trade Discount:

Trade discount means the discount offered by the supplier to the retailer on their purchase value. It is applicable at regular nature at any time. In order to promote the retailer and to earn a certain profit the trade discount is offered. The rate of trade discount is fixed to similar goods by every supplier.

13. Amount accumulated on compound interest basis

$$(A) = P\left(1 + \frac{R}{100}\right)^n$$

In two years a sum becomes Rs 21,632 at a certain rate

$$21632 = P\left(1 + \frac{R}{100}\right)^2 \text{-------- (i)}$$

In three years the sum becomes Rs22497.28 in 3 years

$$22497.28 = P\left(1 + \frac{R}{100}\right)^3 \text{-------- (ii)}$$

Dividing (ii) by (i)

$$1.04 \quad = \quad 1 + \frac{R}{100}$$

$$1.04 - 1 \quad = \quad \frac{R}{100}$$

$0.04 = \dfrac{R}{100}$

R $= 0.04 \times 100$

R $= 4\%.$

$$A = \begin{bmatrix} 5 & -6 & 4 \\ 7 & 4 & -3 \\ 2 & 1 & 6 \end{bmatrix}$$

To find inverse of A we need to find the existence

Type equation here.

$$|A| = \begin{vmatrix} 5 & -6 & 4 \\ 7 & 4 & -3 \\ 2 & 1 & 6 \end{vmatrix} = 5\begin{vmatrix} 4 & -3 \\ 1 & 6 \end{vmatrix} + 6\begin{vmatrix} 7 & -3 \\ 2 & 6 \end{vmatrix} + 4\begin{vmatrix} 7 & 4 \\ 2 & 1 \end{vmatrix}$$

$$= 5(24+3) + 6(42+6) + 4(7\text{-}8)$$
$$= 5(27) + 6(48) + 4(\text{-}1)$$
$$= 135 + 288 \text{-} 4$$
$$= 419$$

Here, $|A| \neq 0$ So, A^{-1} is exist.

Now, Cofactors of A

$A_{11} = + (24+3) = 27$ $\qquad\qquad$ $A_{12} = -(42+6) = -48$
$A_{13} = +(7\text{-}8) = \text{-}1$

$A_{21} = - (\text{-}36\text{-}4) = 40$ $\qquad\qquad$ $A_{22} = +(30\text{-}8) = 22$
$A_{23} = -(5+12) = \text{-}17$

$A_{31} = +(18\text{-}16) = 2$ $\qquad\qquad$ $A_{32} = -(\text{-}15\text{-}28) = 43$
$A_{33} = +(20+42) = 62$

adj. A = *Transpose of* $\begin{bmatrix} 27 & -48 & -1 \\ 40 & 22 & -17 \\ 2 & 43 & 62 \end{bmatrix}$

$$\text{adj.A} = \begin{bmatrix} 27 & 40 & 2 \\ -48 & 22 & 43 \\ -1 & -17 & 62 \end{bmatrix}$$

$$A^{-1} = \dfrac{1}{|A|}(adj.\,A)$$

$$= \frac{1}{419} \begin{bmatrix} 27 & 40 & 2 \\ -48 & 22 & 43 \\ -1 & -17 & 62 \end{bmatrix}$$

$$= \begin{bmatrix} 27/419 & 40/419 & 2/419 \\ -48/419 & 22/419 & 43/419 \\ -1/419 & -17/419 & 62/419 \end{bmatrix}$$

14. **Continuous Variable:**

If two or more variable interlink with each other is called continuous variable. This variable makes chain relation. For example (Price, Demand and Supply), Age, Height and weight of a person) If x, y and z are continuous variable then it existence with continue proportion and this can be expressed as $\frac{x}{y} = \frac{y}{z}$.

Constant:

In any function a fixed portion irrespective of the changes in the variable is called constant. For example in a production function fixed cost is constant like Land, Building and Machine etc.

Explicit function:

Functions with two variables which exist with clear define and separated as independent and dependent variable is called explicit function. For example f(x) =y or f(y) =x.

15. **Commission:**

A commission is revenue received by anyone fulfilling the fixed target. It received on the achievement on turnover with fixed percentages. In business like manager commission, agent commission, dealership commission and sales commission are common. A manager received his commission on his earned profit or sales. Agents are received commission on his sales units under his agency. Dealers are working on basis of fixed percentage of commission in particular of goods they deal.

Brokerage:

The brokerage is income made by the brokers whom intermediary of two parties which they deal and promise to complete the task of financial transaction like arranging assets, securities ,machinery or any support services. It is charged at a fixed percentage on value of his transactions and he avail this from both parties.

16. Percentage is technique of mathematics to express any relation out of hundreds to easier understand and compare. It is important tools of mathematics used in various analyses in commercial, Statistical and scientifically.

    For example if seller's revenue is Rs 50,000 and his earned profit is Rs 10,000 we can find his % of profit. % profit $=\frac{10000}{50000}$ x100 = 20%.So his profit rate in the business is 20%.it makes clear understand to estimate at any level of sales.

17. **Matrices:**

    Matrices are arranging the number in a rectangular form with any row and column. It used in the study of multivariable in economic and scientific fields. Matrices does not hold a single real value, this is expressed in any order of rows and columns.

Determinants:

Determinants are extension of matrices. It is exist only in square matrices and not in any other rectangular matrices. It can be expresses a single real value.

For example $A = \begin{bmatrix} 2 & -5 \\ 3 & 7 \end{bmatrix}$, $B = \begin{bmatrix} 1 & 3 & -2 \\ 7 & 5 & 0 \end{bmatrix}$ are matrices of 2x2 order and 2x3 order in which matrix A has value of determinants by B does not exists the value of determinants since it is not a square matrix.

18. Types of function; refer unit 5.Write any 10 functions briefly for 20 marks.

19. A has 7 years to 20 and B has 4 years to 20.If he divides Rs 40000 as x and 40000-x in the name of A and B respectively,

At compound interest at 9% P.a, both received same sum at the age of 20.

Amount on compound interest $A = P\left(1 + \frac{R}{100}\right)^n$

$x\left(1 + \frac{9}{100}\right)^7 = (40000\text{-}x)\left(1 + \frac{9}{100}\right)^4$

Dividing $\left(1 + \frac{9}{100}\right)^4$ for both side

$x\left(1 + \frac{9}{100}\right)^3 = 40000\text{-}x$

or, 1.295029x +x =40000

or, 2.295029x = 40000

or, x= 40000÷ 2.295029

= 17428.97 or 17430 (approx)

So, the two different investment amounts are Rs 17430 and Rs 22570.

Solution to Business Mathematics
Pondicherry University
B.Com Degree Exam April–2016

Section: A

1. Ratio is a simplest fraction derived from the comparison of similar character or nature of any quantitative measurable things. So, it converts the complicities terms into simpler term to compare or study for any necessities.

2. Joint variations are two or more variables which are interrelated to each other, changes of one leads to changes in other variables are called joined variation.

3. Commissions are the revenue in which Manager or sales Agents received fixed percentages on their sales target or profit.

4. 4:10 or 6:7

$$4:10 = \frac{4}{10} = \frac{4x7}{10x7} = \frac{28}{70}$$

$$6:7 = \frac{6}{7} = \frac{6x10}{7x10} = \frac{60}{70}$$

By the comparison $\frac{60}{70} > \frac{28}{70}$

Therefore 6:7> 4:10

5. Amount invested Rs 5000,Rate of interest(R) =4 ½ %, Time period(T)=1 ½ years

Simple interest $= \frac{PXRXT}{100}$

$$= \frac{5000X\,4\frac{1}{2}\,X1\frac{1}{2}}{100}$$

$$=\frac{5000 \; X \; 9 \; X \; 3}{100 X 2 X 2}$$

= Rs 337.50

6. Matrices are arranging the number in a rectangular form with any row and column. It used in the study of multivariable in economic and scientific fields. Matrices does not hold a single real value, this is expressed in any order of rows and columns.

7. Dividend is distribution of profit to the shareholder by a company from their annual profit. It can be distributed half yearly or annually according to the excess reserve availabity of the company.

8. Polynomial function is a algebraic function in which the exponents of variables are expressed in positive integer number. If the exponent or powers of any variable are negative integers or rational number then it is not a polynomial.

9. Production function is the relation between a firm's production (output) and the material factors of production (input).

10. The equation with degree one and one or two variable is called linear equation. The geometrical representation of the linear equation is a straight line.

Section B

11. a) $2:10 = \frac{2}{12} : \frac{10}{12} = 1.67:8.33$

    b) $100:50 = \frac{100}{150} : \frac{50}{150} = 0.66 : 0.333$

12. Let the share of C = x

Share of B = $\frac{1}{4}$ of C = = $\frac{1}{4}$x

Share of A = $\frac{2}{3}$ of B == $\frac{2}{3}$ of ($\frac{1}{4}$x) = $\frac{1}{6}$x

Now, x+$\frac{1}{4}$x +$\frac{1}{6}$x = 510

x(1+1/4 +1/6) =510

x($\frac{17}{12}$) =510

x $=\frac{510 \times 12}{17}$ = 360

Therefore, share of C = Rs 360, B= Rs 90, A =Rs 60.

13. Answer refers Unit VI Matrices and Determinants.

14. $\int \frac{4x^7-3x^3-5x^2}{x^4}$ dx   (As per the new syllabus this question is irrelevant)

$=\int(4x^3 - \frac{3}{x} -\frac{5}{x^2})$ dx Using the formula $\int x^n$ dx

$=\frac{x^{n+1}}{n+1}$ +c

$=4x\frac{x^4}{4}$ - 3 logx -5$\frac{x^{-1}}{-1}$ +c

$=x^4-3logx +\frac{5}{x}$ +c

15. $\begin{bmatrix} 4 & 1 & 3 \\ 2 & 0 & -6 \\ 5 & -7 & 9 \end{bmatrix} \times \begin{bmatrix} 5 & -2 & 0 \\ 1 & 6 & 8 \\ 3 & 4 & 7 \end{bmatrix}$

$=\begin{bmatrix} (4)5+(1)(1)+(3)(3) & 4(-2)+(1)(6)+3(4) & 4(0)+1(8)+3(7) \\ 2(5)+0(1)+(-6)(3) & 2(-2)+0(6)+(-6)(4) & 2(0)+0(8)+(-6)(4) \\ 5(5)+(-7)(1)+9(3) & 5(-2)+(-7)(6)+9(4) & 5(0)+(-7)(8)+9(7) \end{bmatrix}$

$=\begin{bmatrix} 30 & 10 & 29 \\ -8 & -28 & -24 \\ 45 & -16 & 7 \end{bmatrix}$

16. Amount on compound interest $A = P\left(1 + \dfrac{R}{100}\right)^{n}$

Let the sum invested in two different rate assume Rs x

It gives Rs 21632 in 2 years and 22498 in 3 years

$\therefore x\left(1 + \dfrac{R}{100}\right)^{2} = 21632$ -------- (i)

Also, $x\left(1 + \dfrac{R}{100}\right)^{3} = 22498$ -------- (ii)

Dividing (ii) by (i)

We get, $x\left(1 + \dfrac{R}{100}\right)^{3} / x\left(1 + \dfrac{R}{100}\right)^{2}$
$= 22498/21632$

$\Rightarrow 1 + \dfrac{R}{100} = \dfrac{22498}{21632}$

$\Rightarrow \dfrac{R}{100} = \dfrac{22498}{21632} - 1$

$\Rightarrow \dfrac{R}{100} = \dfrac{22498 - 21632}{21632}$

$\Rightarrow \dfrac{R}{100} = \dfrac{866}{21632}$

$\Rightarrow$ R=4% (Approx)

Putting R=4% in equation (i)

$x\left(1 + \dfrac{4}{100}\right)^{2} = 21632$

$x = 21632 \times \dfrac{100 \times 100}{104 \times 104}$

x=130000

So, the rate of interest is 4% and sum invested Rs 130000.

17. Answers refer Unit-V (write any 6 function with example).

18. $A = \begin{bmatrix} 1 & 0 & -1 \\ 3 & 4 & -5 \\ 0 & -6 & -7 \end{bmatrix}$

To find inverse of A we need to find the existence

Type equation here.

$|A| = \begin{vmatrix} 1 & 0 & -1 \\ 3 & 4 & -5 \\ 0 & -6 & -7 \end{vmatrix}$

$= 1 \begin{vmatrix} 4 & -5 \\ -6 & -7 \end{vmatrix} + 0 \begin{vmatrix} 3 & -5 \\ 0 & -7 \end{vmatrix} - 1 \begin{vmatrix} 3 & 4 \\ 0 & -6 \end{vmatrix}$

$= 1(-28-30) + 0 - 1(-18-0)$

$= 1(-58) + 0 - 1(-18)$

$= -58 + 18$

$= -40$

Here, $|A| \neq 0$ So, A^{-1} is exist.

Now, Cofactors of A

$A_{11} = -48$	$A_{12} = 21$	$A_{13} = -18$
$A_{21} = 6$	$A_{22} = -7$	$A_{23} = 6$
$A_{31} = -4$	$A_{32} = 2$	$A_{33} = 4$

$\text{Adj.A} = \begin{bmatrix} -48 & 21 & -18 \\ 8 & -7 & 6 \\ -4 & 2 & 4 \end{bmatrix}^T$

$\text{adj. A} = \begin{bmatrix} -48 & 8 & -4 \\ 21 & -7 & 2 \\ -18 & 6 & 4 \end{bmatrix}$

$A^{-1} = \frac{1}{|A|} (adj. A)$

$= \frac{1}{-40} \begin{bmatrix} -48 & 8 & -4 \\ 21 & -7 & 2 \\ -18 & 6 & 4 \end{bmatrix}$

$$= \begin{bmatrix} 6/5 & -1/5 & 1/10 \\ -21/40 & 7/40 & -1/20 \\ 9/20 & -3/20 & -1/10 \end{bmatrix}$$

19. The monthly salaries of two person 3:5

    So, the salaries be 3x and 5x

    If the salaries increased Rs 200 the ratio of the new salaries be 13:21

    $\therefore \dfrac{3x+200}{5x+200} = \dfrac{13}{21}$

    $21(3x + 200) = 13(5x+200)$

    or, $63x+4200 = 65x +2600$

    or, $2x = 1600$

    $X = 800$

    $\therefore$ The salaries of two persons are Rs 2400 and Rs 4000

20. Price of a book = Rs 15

    Trade discount = 16%

    Cash discount = 4%

    1 book is free for every 20 books

    Suppose the purchaser buys 20 books

    Cost of 20 books = 20x15 = 300

    Cost Price after trade and cash discount

    = 300-16% of 300 -4% of 300

    = 300-48-12

    = 300-60

    = Rs 240

    $\therefore$ Cost of 21 books = Rs 240

    Cost of one book = 240 ÷ 21

    = Rs 11.43 (Approx)

21. Given equations are

$3x + 2y = 8$ (i)

$5x - 3y = 7$ (ii)

$\Delta = \begin{vmatrix} 3 & 2 \\ 5 & -3 \end{vmatrix} = -9\text{-}10 = -19$

$\Delta x = \begin{vmatrix} 8 & 2 \\ 7 & -3 \end{vmatrix} = -24\text{-}14 = -38$

$\Delta y = \begin{vmatrix} 3 & 8 \\ 5 & 7 \end{vmatrix} = 21\text{-}40 = -19$

$x = \dfrac{\Delta x}{\Delta} = \dfrac{-38}{-19} = 2$

$y = \dfrac{\Delta y}{\Delta} = \dfrac{-19}{-19} = 1$

22. $x_1 + 2x_2 + x_3 = 2$ (i)

    $2x_1 + 2x_3 + x_4 = 6$ (ii)

    $4x_2 + 3x_3 + 2x_4 = -1$ (iii)

    $-x_1 + 6x_2 - x_3 - x_4 = 2$ (iv)

Using Matrix method we can solve the three variables, it is necessary to reduce into three variables.

$x_1 + 2x_2 + x_3 = 2$ (i)

Eqn (ii) x2-Eqn (ii) => $4x_1\text{-}4x_2 + x_3 = 13$ (v)

Eqn (iii) +Eqn (iv) x2=> $-2x_1 + 16x_2 + x_3 = 3$ (vi)

We can be represent the above equations in matrix method as follows,

$$\begin{bmatrix} 1 & 2 & 1 \\ 4 & -4 & 1 \\ 2 & 16 & 1 \end{bmatrix} \begin{bmatrix} x_1 \\ x_2 \\ x_3 \end{bmatrix} = \begin{bmatrix} 2 \\ 13 \\ 3 \end{bmatrix}$$

Let $A X = B$

$$X = A^{-1}B$$

$$|A| = \begin{vmatrix} 1 & 2 & 1 \\ 4 & -4 & 1 \\ 2 & 16 & 1 \end{vmatrix} = 1(-4-16)-2(4-2)+1(64+8)$$

$$=-20-4+72$$

$$= 48$$

Here, $|A| \neq 0$ So, A^{-1} is exist.

Cofactors of A

$A_{11} = -20$	$A_{12} = -2$	$A_{13} = 72$
$A_{21} = 14$	$A_{22} = -1$	$A_{23} = -12$
$A_{31} = 6$	$A_{32} = 3$	$A_{33} = -12$

$$\text{Adj.A} = \begin{bmatrix} -20 & -2 & 72 \\ 14 & -1 & -12 \\ 6 & 3 & -12 \end{bmatrix}^T$$

$$\text{adj. A} = \begin{bmatrix} -20 & 14 & 6 \\ -2 & -1 & 3 \\ 72 & -12 & -12 \end{bmatrix}$$

$$A^{-1} = \frac{1}{|A|}(adj.A)$$

$$= \frac{1}{48}\begin{bmatrix} -20 & 14 & 6 \\ -2 & -1 & 3 \\ 72 & -12 & -12 \end{bmatrix}$$

$$X = A^{-1}B$$

$$= \frac{1}{48}\begin{bmatrix} -20 & 14 & 6 \\ -2 & -1 & 3 \\ 72 & -12 & -12 \end{bmatrix}\begin{bmatrix} 2 \\ 13 \\ 3 \end{bmatrix}$$

$$= \frac{1}{48}\begin{bmatrix} -40 + 182 + 18 \\ -4 - 13 + 9 \\ 144 - 156 - 36 \end{bmatrix}$$

$$=\frac{1}{48}\begin{bmatrix} 160 \\ -8 \\ -48 \end{bmatrix}$$

$$\begin{bmatrix} x_1 \\ x_2 \\ x_3 \end{bmatrix} = \begin{bmatrix} \frac{10}{3} \\ \frac{-1}{6} \\ -1 \end{bmatrix}$$

From the equation (ii) $2(10/3)+2(-1/6)+x_4= 6$

$=> \quad \frac{20}{3} -\frac{1}{3} +x_4 =6$

$=> \quad x_4= 6 -\frac{19}{3}$

$=> \quad x_4=\frac{-1}{3}$